ISBN 978-1-332-79282-5
PIBN 10454258

Bulletin 97

DEPARTMENT OF THE INTERIOR
BUREAU OF MINES
VAN. H. MANNING, Director

SAMPLING AND ANALYZING FLUE GASES

BY

HENRY KREISINGER

AND

F. K. OVITZ

WASHINGTON
GOVERNMENT PRINTING OFFICE
1915

The Bureau of Mines in carrying out one of the provisions of its organic act—to disseminate information concerning investigations made—prints a limited free edition of each of its publications.

When this edition is exhausted, copies may be obtained at cost price only through the Superintendent of Documents, Government Printing Office, Washington, D. C.

The Superintendent of Documents *is not an official of the Bureau of Mines*. His is an entirely separate office and he should be addressed:

SUPERINTENDENT OF DOCUMENTS,
Government Printing Office,
Washington, D. C.

The general law under which publications are distributed prohibits the giving of more than one copy of a publication to one person. The cost of this publication is 15 cents.

First edition. August, 1915.

CONTENTS.

ILLUSTRATIONS.

TABLES.

SAMPLING AND ANALYZING FLUE GASES.

By Henry Kreisinger and F. K. Ovitz.

INTRODUCTION.

Some of the investigations conducted by the Bureau of Mines have for their object the collecting and disseminating of information regarding methods by which the fuels of the country may be most efficiently used. As the analysis of flue gases tends to develop better methods of firing, which in turn reduces waste of fuel, the Bureau of Mines in this bulletin presents for the benefit of those in charge of boiler plants and all other persons interested detailed information on methods of sampling and analyzing flue gases, and on the utilization of the analyses in promoting boiler-room economy.

This bulletin is intended to be a companion to Technical Paper 80,[a] and is written in plain and nontechnical language, as far as possible, so that it may be readily understood by persons who have not had the advantage of a technical education. Whenever possible, illustrations of apparatus and methods of handling have been used rather than elaborate descriptions.

The material presented in this report is arranged in two parts. The first part contains the description of the apparatus and the methods used in sampling and analyzing flue gases. The second part gives experimental results obtained with the different methods of sampling and collecting flue gas that are recommended in the first part of the report.

Only simple apparatus and methods are considered, as they are accurate enough to show the large heat losses due to the use of too much air, and are also accurate enough to indicate any incomplete combustion losses of economic importance. Without doubt the loss due to large excess of air is the greatest one in the boiler room, and can usually be greatly reduced by making proper use of gas analysis. Before engineers in the isolated industrial plants can be induced to analyze for small traces of combustible gases, they must be first taught how to analyze for carbon dioxide (CO_2), and they must learn to appreciate the great possibilities of reducing loss from large excess of air. Furthermore, in the face of the great difficulty of obtaining a fair

---

[a] Kreisinger, Henry, Hand firing soft coal under power-plant boilers: Tech. Paper 80, Bureau of Mines, 1915 83 pp.

average sample, discussed in the second part of this bulletin, it is doubtful whether more delicate apparatus for analysis and more refined methods would be of much advantage. Those who wish to obtain information on the more accurate methods of analysis of gases are referred to Bureau of Mines Bulletin 12[a] and Technical Paper 31.[b]

GENERAL STATEMENT.

The determination of the composition of flue gases involves two distinct operations—the sampling of the gases, and the analysis of the sample. Of the two operations the sampling is more important than the analysis, for if the sampling is not done properly the analysis itself is of no value. Also, sampling is usually more laborious than analysis. Great care is necessary, just as it is necessary in sampling coal. In sampling either flue gas or coal a carelessly taken sample is worthless for determining the quality of the gas or coal. Therefore, the device and the methods used in sampling should be well planned and the sampling itself carefully executed.

THE SAMPLING OF FLUE GASES.

In sampling flue gases a small quantity is taken from the stream of gases at any desirable place within a boiler setting. The sample may be taken from the furnace, from among the boiler tubes, or from the breeching, according to the object of the sampling and analysis. The volume of the sample may range from one-half of a cubic foot to a few cubic inches.

In sampling the flue gases the most important feature is that the sample shall closely represent the composition of the stream of gases at the place of sampling, and the device used in sampling must be designed with this object in view. A simple design of such a device for sampling gases in the uptake of a boiler is shown in figure 1.

It consists mainly of a special piece of piping, called the sampler, inserted in the uptake, two collecting bottles, and a steam ejector. These parts are connected by a small-size standard iron piping or ¼-inch copper tubing, as shown in the figure. The operation of the device is indicated by the arrow heads. The steam ejector causes a continuous flow of gas from the uptake through the piping. A small amount of this gas is drawn into the upper collecting bottle and is the sample to be analyzed. The remainder of the gas is rejected with the steam from the ejector through the discharge pipe.

The ejector and the collecting bottles may be placed in any convenient place. In most plants the engine room is the best location;

---

[a] Frazer, J. C. W., and Hoffman, E. J., Apparatus and methods for the sampling and analysis of furnace gases: Bull. 12, Bureau of Mines, 1911, 22 pp.

[b] Burrell, G. A., and Seibert, F. M., Apparatus for the exact analysis of flue gas: Tech. Paper 31, Bureau of Mines, 1913, 12 pp.

usually they can be placed along the wall separating the engine room from the boiler room. Such location requires a short pipe connection which is always advantageous and easily supervised by the engineer.

Figure 2 shows the location of the ejector, the collecting bottles, and the pipe connections in a typical steam plant. Each boiler is equipped with a separate sampling device independent of the other boilers. One ejector will serve several sampling devices.

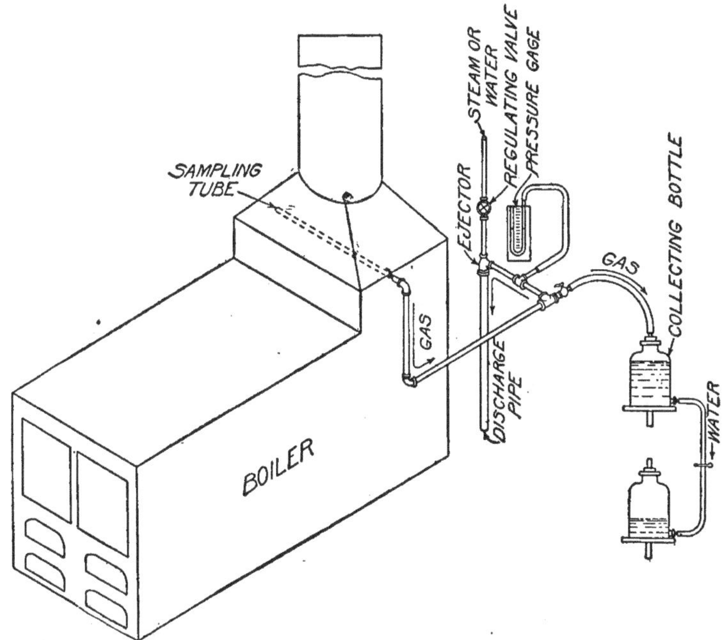

FIGURE 1.—General arrangement of the parts of a flue-gas sampling device.

TYPES OF SAMPLERS.

That part of the sampling device that is inserted into the uptake or into any other place from which it is desired to take a gas sample is called the sampling tube or the sampler. The construction and the location of the sampler are of considerable importance in obtaining a representative gas sample.

SIMPLE OPEN-END SAMPLER.

The simplest and the most commonly used sampler is a plain standard ¼-inch iron pipe placed with the open end in the center of the uptake as shown in figure 3. The gas is taken only from the center of the stream flowing through the uptake, and the sample collected represents only that part of the stream.

The parts of the stream farther away from the center may be and usually are of somewhat different composition than the central part. This difference may amount in some cases to 2 per cent in the CO_2 content; that is, in the central part the gases may contain 8 per cent of CO_2, in another part farther away from the center 10 per cent, and in some other part only 6 per cent. If the setting is very leaky the variation may be greater. In water-tube boilers if the setting is fairly tight the composition of the central part will be within about 0.5 per cent of the average composition of the whole stream. In fire-tube boilers the difference may be as much as 2 per cent. These variations are discussed in the section on "Experimental Results of Sampling and Collecting Flue Gases."

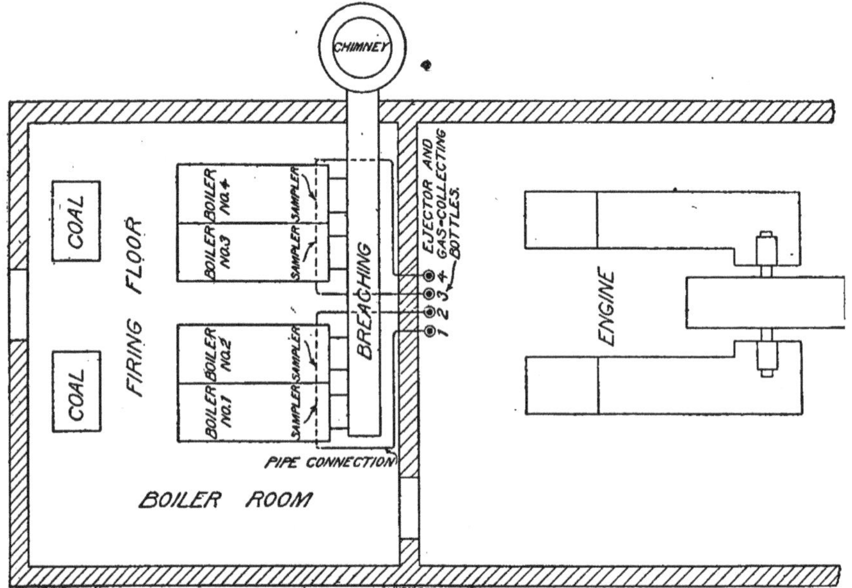

FIGURE 2.—Position of gas-sampling devices and their connections in a typical steam plant. Each of the four boilers is equipped with a separate sampling device.

The open-end single-tube sampler shown in figure 3 has the merit of simplicity. It is easily put in place and can be cleaned readily in case it becomes clogged with soot or ashes. When this sampler is installed the only precaution to be taken is that the open end be placed in the center of the uptake at least 3 feet below the damper and not any nearer than 1 foot to the steam drum. Figure 3 shows a good position for such sampler in a water-tube boiler installation. The sampler is inserted into the uptake through a hole drilled in the brick wall. Leakage around the sampler into the uptake can be stopped with asbestos packing. Figure 4 shows a good placing of the single-tube sampler in a horizontal tubular boiler.

The sampler is placed in the smoke box about 1 foot above the top row of tubes. If a sample is to be taken from any other place within the boiler setting, it is preferable that the open end of the sampler be placed in the center of the stream of gases.

PERFORATED-TUBE SAMPLER.

When it is desirable to obtain a sample that more nearly represents the composition of the entire stream of gases, a sampler of the type shown in figure 5 can be used.

This sampler consists of standard 1-inch iron pipe long enough to reach across the uptake or smoke box. The pipe is closed at both ends. The gas enters into this pipe through a number of $\frac{1}{16}$-inch holes drilled in opposite sides of the pipe about 6 inches apart, as shown in the figure. Through the bushing closing one end of the pipe is inserted $\frac{1}{4}$-inch iron pipe, which extends to the middle of the 1-inch pipe. This inside pipe is connected to the ejector and draws the gas out of the sampler. Such a sampler takes the gas the full width of the uptake at nearly uniform rate, and therefore the sample collected with it more nearly represents the entire stream of gas. Its drawback is that occasionally the small holes become stopped with soot and fine ashes. On this account the sampler must be taken out every few weeks and cleaned. Ordinarily if the gas analysis is used only as an aid to running the fires, it is doubtful whether the extra trouble of making this sampler and keeping it clean pays for the greater accuracy. If, however, the gas analysis is used for computing chimney losses the perforated sampler should be used, especially with fire-tube boilers.

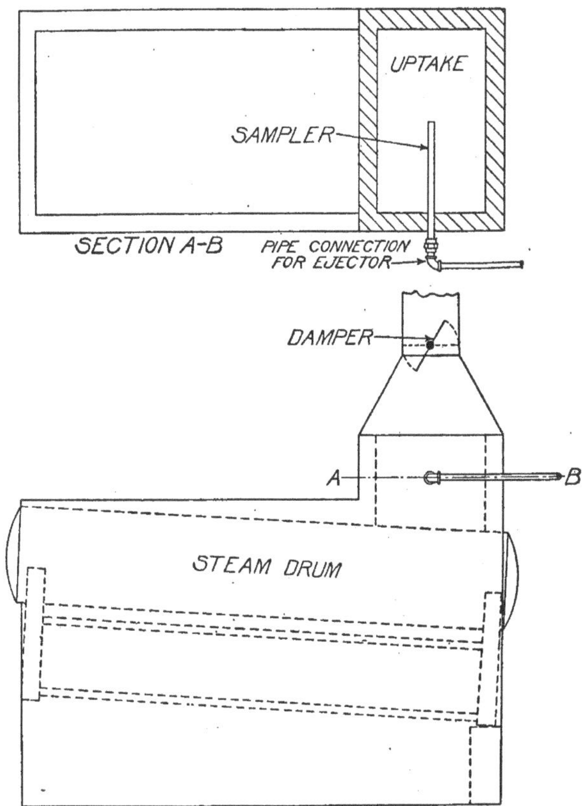

FIGURE 3.—Simple open-end sampler placed in the uptake of a water-tube boiler.

PIPE CONNECTIONS FOR SAMPLER.

The pipe connection is easiest made of standard ¼-inch iron pipe and cast-iron fittings. If possible the pipe should be so laid that there is a gradual drop toward the ejector. There should be no "pockets" in the pipe line, as these would fill up with condensed water vapor, which is always contained in the flue gases. If it is necessary

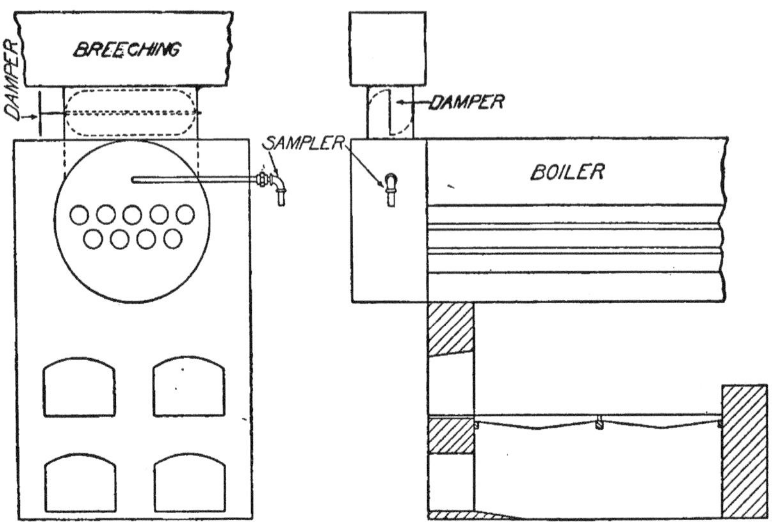

FIGURE 4.—Simple open-end sampler placed in the uptake of a horizontal return-tubular boiler.

to lower the pipe line at any place it is advisable that each low place be provided with a drain cock and a water trap, as shown in figure 6.

During the collection of a sample the drain cock is kept open, and the condensed moisture drains into the bottle. When the bottle is nearly filled, the drain cock can be closed and the bottle emptied. The connections on the water trap should be carefully put together, so that no air can leak into the pipe line. A rubber stopper should

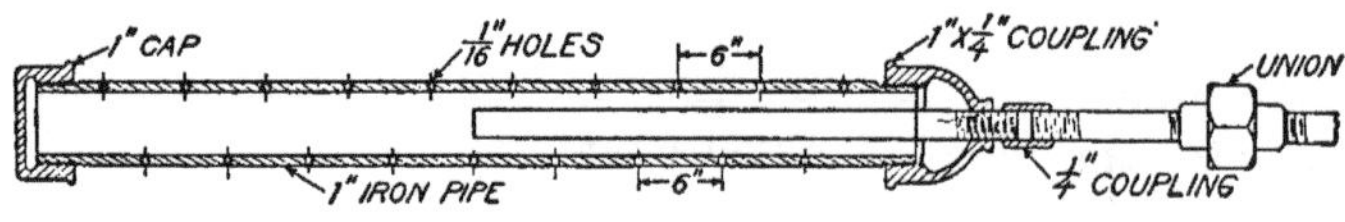

FIGURE 5.—Perforated-tube sampler.

be used in the flask in which the water collects; cork stoppers can not be kept air-tight.

The kind of fittings used at the ejector end of the pipe connection and the arrangement of these fittings are shown in figure 7.

The cocks best suited for the purpose are ¼-inch brass cocks, with male connection, corrugated hose end, and lever handle. Such cocks can be purchased from any dealer in gas fittings for about 20 to 25 cents apiece. The corrugated hose end makes good connection with a ¼-inch rubber tubing. It is important that the cock delivering gas to the collecting bottles should be connected directly to the gas line as shown in figure 7; that is, no piping should be placed between the cock and the T in the gas line. Such piping would form a pocket of stagnant gas or air which would be drawn into the collecting bottle when the collection of a sample was started. All the joints should be well put together with graphite compound and painted with some thick paint to reduce as much as possible the chances of leakage. Rubber connections should not be painted.

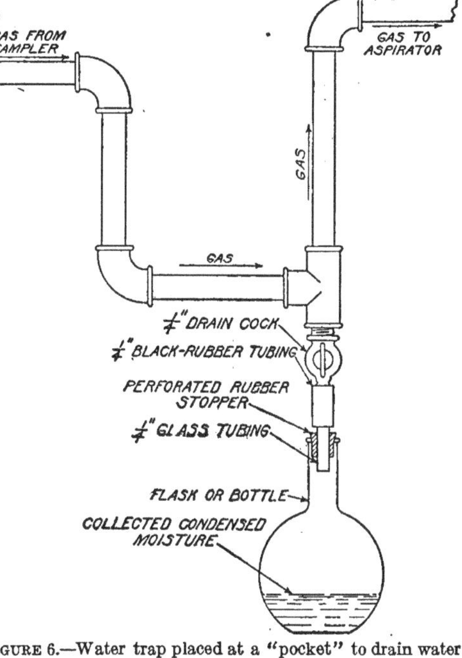

FIGURE 6.—Water trap placed at a "pocket" to drain water condensation.

EJECTOR.

The ejector works on the same principle as the ordinary steam injector used for feeding boilers. An efficient ejector can be made of ¼-inch fitting, as shown in figure 7. The only part that must be made specially is the nozzle. The nozzle consists of a ¼-inch nipple about 1½ inches long, made out of a solid piece of ½-inch round brass, with $\frac{1}{16}$-inch hole. The threading of the ends of the nipple and the drilling of the $\frac{1}{16}$-inch hole had better be done in the lathe, so as to have the hole in the center and the threading true with the hole. It is advisable not to drill the hole any larger than one-sixteenth inch,

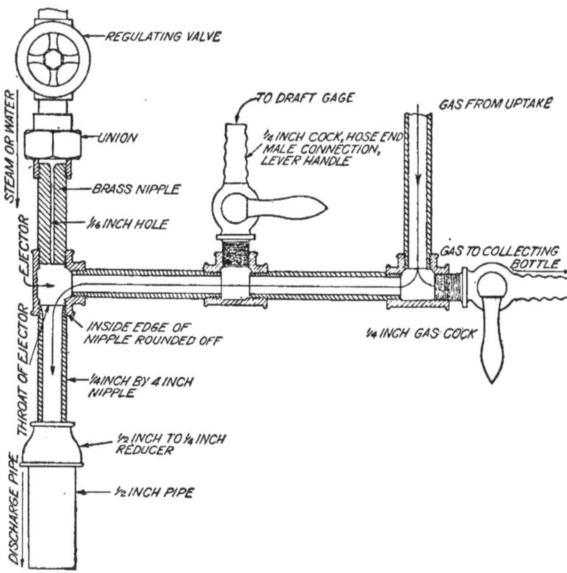

FIGURE 7.—Details of ejector and connections.

because too much steam would be used and the discharge pipe would have to be made of a large-size pipe; otherwise the steam will build up back pressure and reduce the vacuum caused by the ejector. The ¼-inch nipple forming the throat of the ejector should have the inside edge well rounded off, as shown in the figure. With this edge rounded off, the ejector operates more steadily and higher vacuum can be obtained.

In putting the ejector together care should be taken that the nozzle is threaded into the T straight, so that the stream of water or steam issuing from the nozzle hits the center of the throat. If the stream hits too far off the center the suction effect of the ejector might be thereby greatly reduced. With a well-made ejector, using steam, a suction sufficient to lift 3 or 4 feet of water column can be obtained. The suction can be adjusted to any lower value by a steam-regulating valve. A suction of 3 or 4 inches of water is usually sufficient. If water under about 50 pounds' pressure is used in the ejector, a suction of about 5 inches of water can be obtained. When water is used, the ¼-inch pipe forming the throat of the injector should be at least 12 inches long. If steam is used, the discharge pipe beyond the throat should be short and with as few elbows in it as possible. A union placed between the nozzle of the ejector and the regulating valve facilitates cleaning of the nozzle.

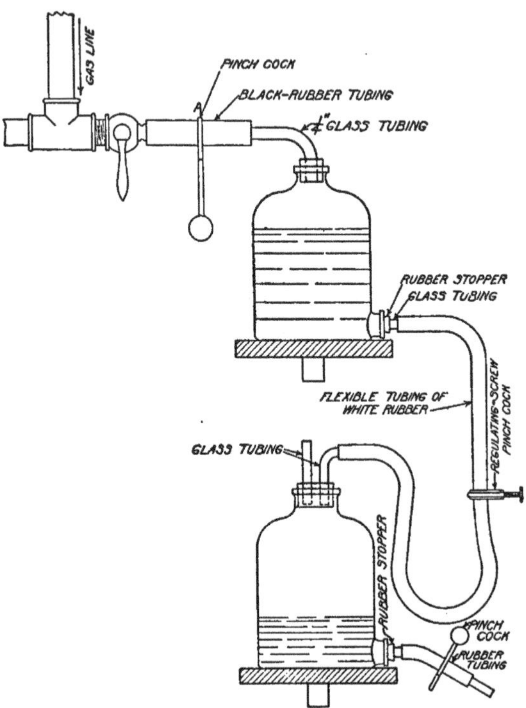

FIGURE 8.—Collecting bottles, showing their connections and arrangement when collecting gas.

COLLECTING BOTTLES AND THEIR CONNECTIONS.

The collecting bottles and their connections are shown in figure 8. Their trade name is "aspirator bottles." They have a side opening near the bottom. For collecting samples over a period of half a day bottles of 2 gallons capacity each are advisable. They can be purchased from any chemical supply or scientific-apparatus supply company at a price of about $1.50 to $2 each.

Both openings of each bottle are provided with well-fitting rubber stoppers, which should be ordered with the bottle. The stopper in the mouth of the bottle has two $\frac{1}{4}$-inch perforations; the stopper in the side opening has only one $\frac{1}{4}$-inch perforation. Cork stoppers should not be used, because they can not be kept gas-tight.

GLASS TUBING.

The glass tubing used in making the connections is $\frac{9}{32}$-inch outside and about $\frac{3}{16}$-inch inside diameter. About 20 feet of the tubing weighs 1 pound. Soft glass is better suited for this purpose, because the tube can be easily bent to any shape by heating to a dull-red temperature in a gasoline torch. While the tubing is being heated, it should be slowly revolved in the flame, so that it is heated uniformly around the circumference. If a bend at right angles is to be made, about 2 inches of the tube should be heated to a uniform dull-red heat and then removed from the flame before attempt is made to bend it. A wide flame is preferable to a pointed one for heating glass tubing that is to be bent. If the flame can not be made wide enough, the tubing should be moved back and forth sideways, so as to get sufficient length of the tubing heated. If only a short piece of the tubing is heated, a very sharp bend will be formed and the tubing will flatten down and nearly close the hole.

The tubing can be broken off at any desired place by filing a deep scratch on the glass and then exerting a slight pressure with the thumbs on the side opposite the scratch.

The glass tubing can be purchased from any chemical supply company at about 40 or 50 cents a pound. It comes in lengths of about 30 to 36 inches.

RUBBER TUBING.

The rubber connection between the upper bottle and the gas line should be made of $\frac{1}{4}$-inch black seamless unvulcanized rubber tubing of heavy walls. Such tubing can be purchased from any chemical supply company for about 15 cents a foot. The black rubber remains elastic and maintains gas-tight joints, and should be used wherever a gas-tight joint is required.

The connection between the two bottles can be of $\frac{1}{4}$-inch white-rubber tubing, light wall, cloth impression. Such tubing can be purchased for about 8 or 9 cents a foot. It is not so elastic as the black-rubber tubing and therefore it is not recommended for gas connections that must remain tight. However, for the water connection between the two bottles it is cheaper and even more durable than the black tubing.

TUBING CLAMPS.

Three clamps are needed for each set of collecting bottles. Of these the one used on the water connection between the two bottles should be a screw clamp (fig. 9). This type of clamp permits an easy adjustment of the quantity of water flowing from the upper bottle into the lower one. The other two clamps should preferably be of the spring type commonly called "pinch cocks" (fig. 9). This type can be used to better advantage where it is desired to have the rubber connection fully open or completely closed.

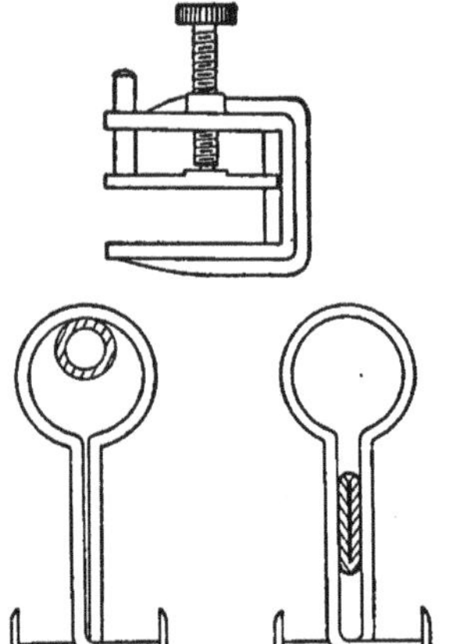

FIGURE 9.—Tubing clamps for closing rubber tubing. Top view shows a screw clamp to be used when fine adjustment of the flow through the tube is desired. Bottom views show a pinch cock for completely closing the rubber tubing.

The cost of the large screw clamp is about 15 cents; the price of the spring pinch cocks is about 6 cents each. They can be bought from any chemical supply company. The screw pinch cock should have the end of the screw covered with metal, so that the screw will not cut the rubber tubing.

PROCEDURE IN COLLECTION OF GAS SAMPLES.

DETAILS OF ARRANGING APPARATUS.

Figure 8 shows the arrangement of the bottles with connections made for collecting a gas sample. The upper bottle is connected to the corrugated hose end of the lever cock. This connection should be as short as possible. The lower bottle is placed 2 to 4 feet below the upper one. The greater this difference between their elevations, the more uniform is the rate of collecting the gases over long periods. The rubber-tube connection enters the lower bottle through the mouth and not through the side opening.

At the start the lower bottle is nearly empty. The upper one is completely filled with water; in fact, it is advisable to run the water into the rubber tubing as high as the pinch cock, so that all the gas or air will be expelled from the bottle. Care should be taken that the short piece of glass tubing does not protrude below the rubber stopper. If it should, a pocket of gas will be formed below the stopper and it will be impossible to expel this gas. All the pinch cocks as well as the lever cock on the gas line are shut. The ejector should run several minutes with a suction of 3 or 4 inches of water before the collection of the gas sample is begun. The suction is measured by the draft gage attached as shown.

To start collecting the gases the lever cock on the gas line and the pinch cock *A* are fully opened. The regulating screw pinch cock is then partly opened and adjusted to any desired flow of water from the upper bottle into the lower one. The correct adjustment of the regulating cock can be made much easier by observation of the quantity of water dripping into the lower bottle than by observing the dropping of the water level in the upper bottle. When the water drips at the rate of three drops in 2 seconds, it takes about 6 hours to empty the upper bottle. As the water flows out of the upper bottle gas is drawn in from the gas line and takes the place of the water. The air in the lower bottle escapes through the short piece of glass tubing in the rubber stopper.

METHOD OF OBTAINING EFFECTIVE HEAD OF WATER.

The effective head causing the water to flow through the rubber connection between the two bottles is the difference between the elevation of the water level in the upper bottle and the elevation of the mouth of the lower bottle. This effective head is changed only by the drop of the water level in the upper bottle and not by the rise of the water level in the lower one. This relation holds only when the bottles are connected as shown in figure 8.

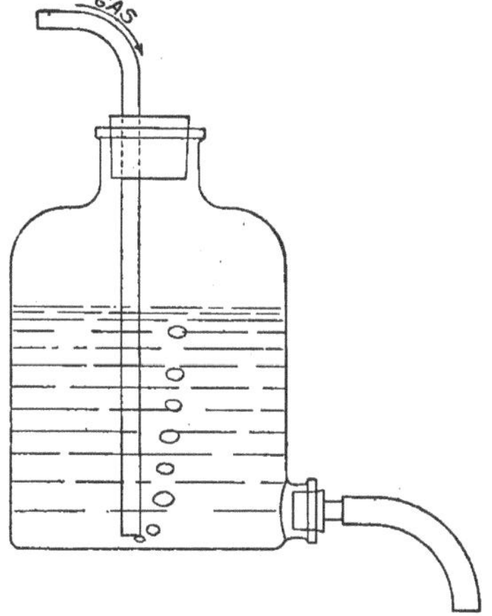

FIGURE 10.—Connections of the upper collecting bottle to maintain constant effective head. The gas bubbles through the water. This arrangement is not recommended.

If the connections were made to the side outlet of the lower bottle, the rise of the water level in it would also diminish the effective head. Variation in the effective head causes a corresponding variation in the rate of flow of water from the upper bottle into the lower one. Consequently the rate of collecting the gas sample is reduced as the effective head becomes smaller, although the adjustment of the regulating pinch cock remains the same. Therefore the variation in the effective head should be as small as possible. The greater the difference in the elevation of the two bottles the smaller will be the reduction in the rate of collecting the gas sample due to the drop of water level in the upper bottle. A difference of more than 2 feet gives fair results.

A constant effective head can be obtained by extending the glass tube so as to bring the gas nearly to the bottom of the upper bottle, and so as to let the gas bubble through the water, as shown in figure 10.

With this arrangement the effective water head is equal to the difference of elevation between the lower end of the glass tube and the mouth of the lower bottle and is of course constant. However, this arrangement is not recommended, because as the gas bubbles through the water a considerable amount of the carbon dioxide (CO_2) may be absorbed by the water or given off, and an error amounting to several per cent may be introduced in the CO_2 determination. The seriousness of this error is shown by experimental results given in Table 9 on page 63.

EFFECT OF ABSORPTION OF CARBON DIOXIDE IN WATER.

Carbon dioxide is much more soluble in water than oxygen and nitrogen. The respective solubility of these gases is approximately 90, 2, and 1. Furthermore, water from a well or river is usually more nearly saturated with oxygen and nitrogen than with carbon dioxide. When the upper collecting bottle is freshly filled with water and flue gas is bubbled through the water, as shown in figure 10, a large part of the carbon dioxide is absorbed by the water. Thus if the flue gas contains 12 per cent of CO_2 when it enters the bottle, 6 per cent of the 12 may be absorbed by the water, so that the analysis of the collected sample may show only 6 per cent of CO_2. If several consecutive samples containing 12 per cent of CO_2 are collected, the water becomes saturated and the absorption of CO_2 is gradually reduced. Finally, the analysis of the last sample may show nearly 12 per cent. If, however, after the water has been saturated with flue gas containing 12 per cent of CO_2, a sample having only 6 per cent of CO_2 is collected, the water gives off some of the CO_2 absorbed from previous samples, and the analysis of the sample may show 9 per cent of CO_2.

That the error caused by the absorption of CO_2 by water is a serious one is shown by the experimental results given on pages 59 to 64. Even when the gas is collected over water, as shown in figure 8, some carbon dioxide may be absorbed or given off, although to a smaller degree. The absorption is larger if the gas sample is allowed to stand in the bottle a long time. Therefore, to reduce this error, the sample of gas should be analyzed immediately after the collection has been completed. The error due to this cause can be further reduced if salt brine be used instead of water. Carbon dioxide is less soluble in brine than in water. The absorption of CO_2 could be completely avoided if the bottles were filled with mercury instead of brine or water. However, the cost of mercury and the increased difficulty in handling the apparatus make the use of mercury inadvisable for the ordinary boiler-room or engine-room work.

DISPOSAL OF SURPLUS OF SAMPLE.

After a sample has been collected and a part of it taken for analysis, the remainder of the gas can be discharged into the gas line by raising the lower bottle and changing the connection to its side outlet, as shown in figure 11. All the cocks, including the lever cock on the gas line, are fully opened.

The water flows back into the bottle connected to the gas line and forces the gas out through the discharge line. When the water has risen to the pinch cock *A*, all the cocks are closed and the bottles changed to the position shown in figure 8. The bottles are then ready for collecting another sample.

COLLECTION OF "GRAB" SAMPLE WITH ORSAT APPARATUS.

Sometimes it is desired to collect a sample of gas from various places within the setting over a short period of a few seconds. This can be done most easily by drawing the gas sample directly into the measuring burette of an Orsat apparatus, using an open-end tube sampler. Figure 12 shows a method of taking a "grab" sample from the third pass of a cross-flow water-tube boiler, the sampling tube being inserted through the cleaning holes.

Each time a sample for analysis is taken, the air should be removed from the sampler and the connections by drawing the air into the measuring burette and discharging it through the three-way cock.

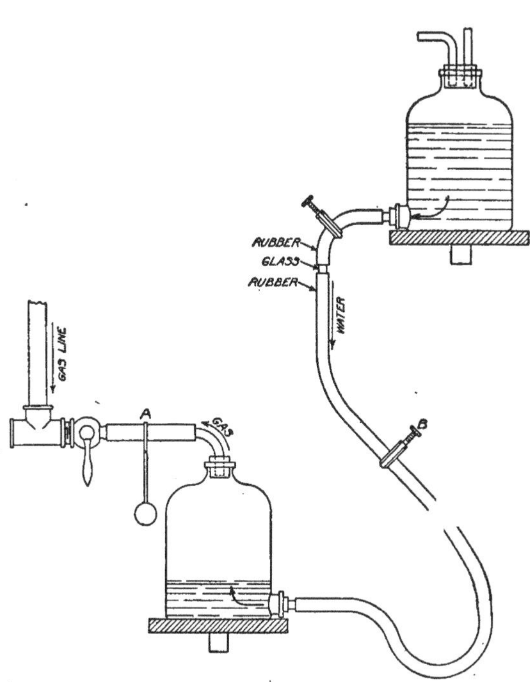

FIGURE 11.—Position of bottles when unused part of gas sample is being discharged and the bottle used for collecting sample is being filled with water. *A*, spring pinch cock; *B*, screw pinch cock.

The sampler should be made of small-size tubing, and the rubber connection should be as short as practicable in order that the volume of air they hold may be small and quickly drawn out. Ordinarily, filling the measuring burette two or three times suffices to remove the air. If only CO_2 is to be determined, $\frac{3}{16}$ inch or $\frac{1}{4}$ inch copper tubing is convenient for use as a sampler. One-quarter inch copper tubing holds approximately 1 cubic inch of gas for every 3 feet of its length, so that the measuring burette of a half-size Orsat apparatus at one filling empties about 9 feet of the tube. The capacity of $\frac{1}{4}$-inch standard iron pipe is about 1 cubic inch of gas for every 10 inches of its length, so that one filling of the measuring burette of a half-size apparatus empties only about 30 inches of the pipe. These figures show the advantage of using small-size tubing as a sampler when the gas is collected directly in the Orsat apparatus.

The perforated-pipe sampler shown in figure 5 is not well adapted for this method of sampling because of its large gas-holding capacity. The manipulation of the Orsat apparatus when the sample is being drawn is described in the next chapter.

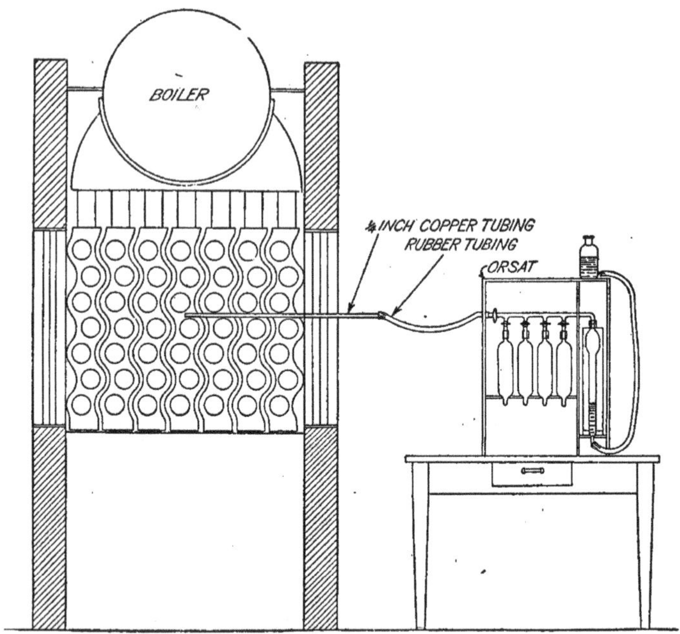

FIGURE 12.—Arrangement of apparatus for collecting "grab" sample from the third pass of a vertically baffled water-tube boiler.

ANALYSIS OF FLUE GASES.

Flue gas consists of carbon dioxide, oxygen, carbon monoxide, and nitrogen; but sometimes may contain small proportions of hydrogen and methane. Water vapor and a little sulphur dioxide are always present, but as regards boiler-room economy these constituents have little significance. Usually only the carbon dioxide, oxygen, and carbon monoxide are determined and the remainder is considered nitrogen. Frequently the carbon dioxide is the only constituent determined, this alone giving sufficient information as to the economic performance of the furnace.

The flue gases are usually analyzed with a portable apparatus commonly known as the Orsat, shown in figures 13 and 14. An analysis consists of successively absorbing the different constituents of the gas sample and of measuring the volume of the sample before and after each absorption.

DESCRIPTION OF ORSAT APPARATUS.

The essential parts of the Orsat apparatus are a measuring burette, three absorption pipettes, a leveling bottle, and a header, all of which are made of glass. These glass parts are inclosed in a compact wooden case.

MEASURING BURETTE.

The measuring burette is a cylindrical glass vessel graduated into 100 units, each unit being graduated into fifths. The measuring burette of a full-size Orsat apparatus has a volume of 100 c.c.; the volume of the burette of the half-size apparatus is 50 c. c. A volume of 16.38 c.c. is equal to 1 cubic inch. The burette is used for measuring the volume of gases during the process of analysis. It is inclosed in a water jacket which prevents sudden changes in temperature while the analysis is being made.

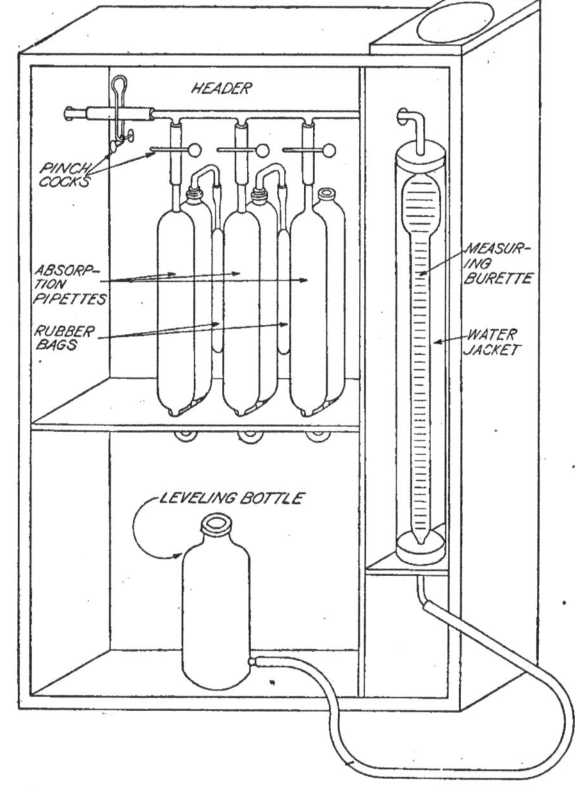

FIGURE 13.—Orsat apparatus with wire pinch cocks and rubber bags.

ABSORPTION PIPETTES.

The absorption pipettes are U-shaped glass vessels and contain solutions for absorbing the three principal constituents of the flue gas. The one nearest to the measuring burette contains potassium hyrdoxide solution for absorbing carbon dioxide; the second one contains an alkaline solution of pyrogallic acid for absorbing oxygen; and the third contains ammoniacal solution of cuprous chloride for absorbing carbon monoxide.

Two kinds of absorption pipettes are shown in figure 15.

The pipette with the opening and the ground-glass stopper near the top is not as good a design for general boiler-room work as the plain one. The stopper may become loose and the gas or solutions may leak out and cause trouble. One arm of the pipette contains glass tubes to increase the surface of the liquid when in contact with the gas. To the other arm is attached a water seal or rubber bag to protect the solutions from deterioration by contact with air. The method of attaching the water seal or rubber bags is shown in figure 16.

The water seal requires an extra pipette, three No. 0 rubber stoppers with one hole, a glass T, and two right-angled glass bends. The joints should be made with ¼-inch black-rubber tubing. The rubber bags are attached by means of a rubber stopper and short piece of glass tubing. They should be air tight, as otherwise they may cause a great deal of trouble. After two or three months the rubber becomes hard and porous and the bags should be replaced by new ones. The water seal is more difficult to attach, but once attached to the apparatus it is permanent and gives no trouble. Only the solutions for absorbing oxygen and carbon monoxide need to be protected.

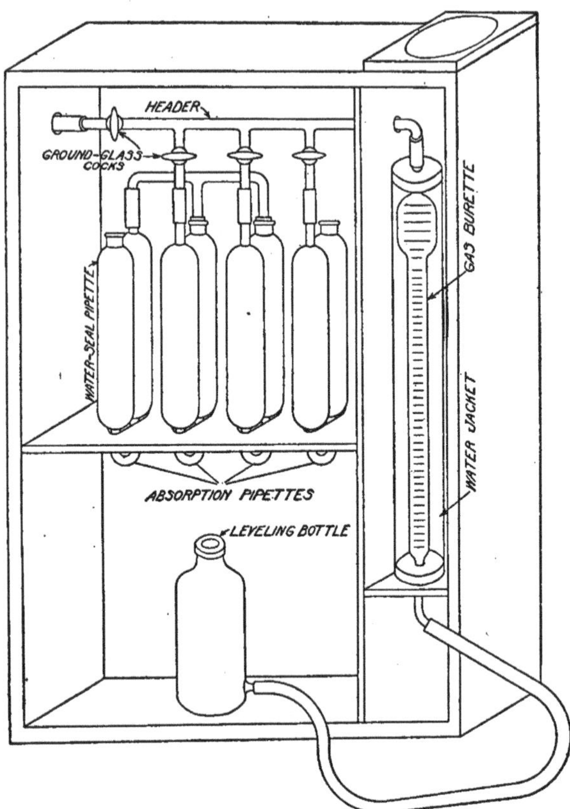

FIGURE 14.—Orsat apparatus with glass stopcocks and water seal.

LEVELING BOTTLE.

The leveling bottle, which is connected to the measuring burette with about 3 feet of black-rubber tubing, contains water. Lowering or raising the leveling bottle causes the water to act as a piston and draw gas into the measuring burette or force gas out of it into any absorption pipette. The water in the leveling bottle also is used to regulate the pressure on the gas when its volume is being measured.

HEADER.

The header is made of glass tubing of small bore. The measuring burette and absorption pipettes are attached to it with ¼-inch black-rubber tubing. All the joints must be gas tight. There is a glass stopcock on the header between each pipette and the burette, and at one end there is a three-way stopcock through which gas is taken into or discharged from the apparatus. The three-way cock has a spot of colored glass on it to indicate the direction of flow. Spring pinch cocks can replace the three stopcocks between the pipettes and the burette and thereby somewhat reduce the cost of the apparatus. The writers find the glass cocks more convenient to manipulate, but the glass cocks require greater care to keep them in good working condition. They must be kept lubricated with vaseline so that they will turn easily and will not stick. The vaseline should be applied in a thin film on the bearing surface of the valve and all excess avoided, for if too much is applied the small holes in the valves may become stopped.

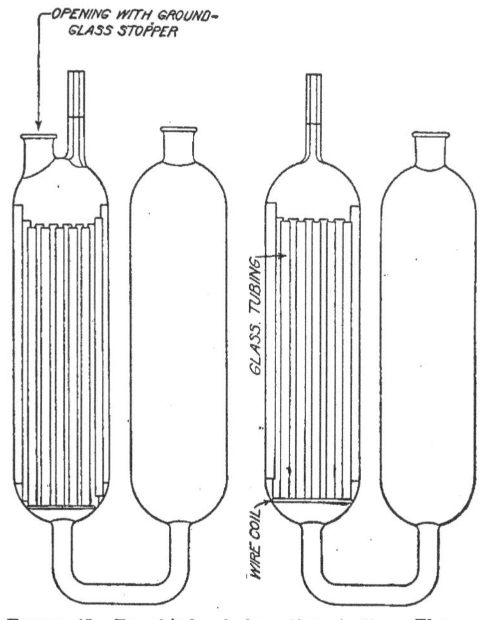

FIGURE 15.—Two kinds of absorption pipettes. The one without ground-glass stopper is the better design.

COST OF APPARATUS.

The half-size Orsat is lighter than the full size and therefore is preferable if the apparatus is to be carried from place to place. The apparatus can be purchased from any chemical-supply company. The cost of the full-size type with glass stopcocks is about $20, or with spring pinch cocks about $18. The half-size type correspondingly equipped costs about $18 and $16. Usually the apparatus is fitted with rubber bags to protect the solutions; if the water seal is wanted it must be specified when purchasing.

MANIPULATION OF ORSAT APPARATUS.

In a gas analysis the manipulation of the Orsat apparatus consists mainly of two operations—moving the gases into the different parts of the Orsat apparatus and measuring their volume.

MOVING THE GASES.

The moving of the gases is done with the water in the measuring burette and the leveling bottle. The water acts as a piston; when the leveling bottle is raised the water flows by gravity into the measuring burette and forces the gases into one of the absorption pipettes or out of the apparatus, depending on which valve is open. If the leveling bottle is lowered, the water flows by gravity from the burette into the leveling bottle and the gas is drawn from the apparatus into the burette. This manipulation in indicated in figures 17 and 18.

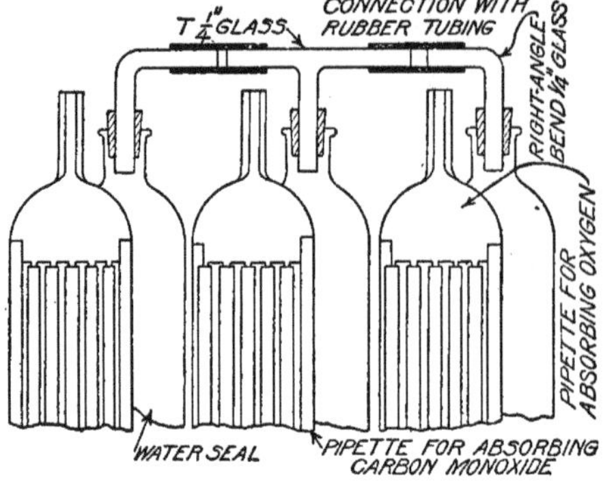

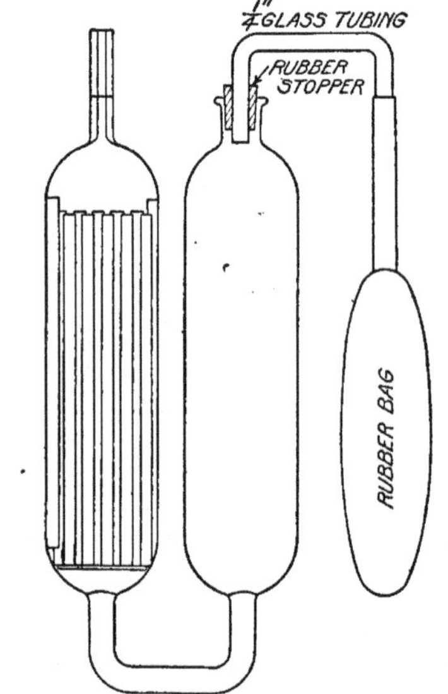

FIGURE 16.—Method of attaching water seal and rubber bags for protection of the solutions from air.

In the illustrations the movement of water and gases is indicated by the arrows. Figure 17 shows the process of drawing the gas sample from the collecting bottle into the Orsat apparatus, the water flowing out of the burette into the leveling bottle. The three-way cock on the Orsat apparatus is opened in such a way that the gases flow from the collecting bottle into the measuring burette; the valves on all the absorption pipettes are closed.

When gas is being drawn from the absorption pipette, care should be taken not to draw the solution into the valve or to run the solution into the water in the measuring burette. The solution rises slowly as long as the level is in the wide part of the pipette, but as soon as it reaches the small bore of the neck it rises rapidly, and before an inexperienced operator realizes what is happening part of his solution is drawn into the measuring burette. The best way to avoid such accidents is to operate the Orsat apparatus as shown in figure 19. The leveling bottle is manipulated with the right hand while the rubber-tube connection is held between the thumb and first finger of the left hand.

If pressure be exerted on the rubber tubing, the flow of gas can be so reduced that it can be stopped easily with the top of the solution at any desired point in the neck of the pipette. When the solution has been brought to the point desired, the analyst may hold it there by pinching the rubber tubing, the leveling bottle being placed on the stand so as to leave the right hand to manipulate the valve. The same manipulation should be used when the gases are being forced into the absorption pipettes.

MEASURING VOLUME OF GASES.

The measurement of the volume is the most important operation in the analysis of gases. During any one analysis all measurements of the gas volume must be made under constant temperature and pressure.

A rise in temperature increases the volume of the gases. It is to avoid the change in the temperature that the measuring burette is water-jacketed. To further decrease the possibility of temperature changes, the Orsat apparatus should be in a place free from draft or heat radiation from the boiler or engine.

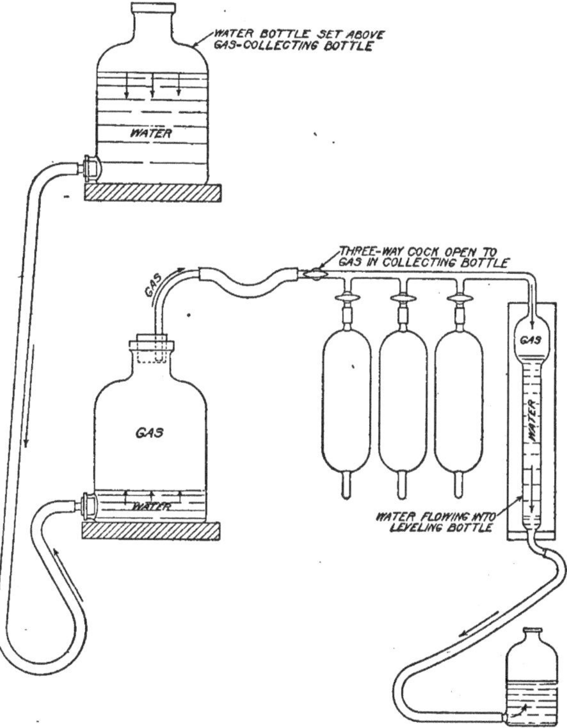

FIGURE 17.—Relative position of collecting bottle, Orsat apparatus, and leveling bottle, and their connections when gas is being drawn into the burette.

A change in pressure changes the volume of gases. Inasmuch as the atmospheric pressure changes little and any change takes place slowly, the volume of gases is conveniently measured under atmospheric pressure. The gas in the measuring burette is brought under atmospheric pressure by holding the leveling bottle at such position that the water level in the measuring burette and that in the leveling bottle are at the same height; that is, so that a horizontal line will pass through both levels. The position of the leveling bottle when the volume of the gases is read is shown in figure 20. When the reading is taken, the rubber tubing must be free from kinks.

The necessity of bringing the two levels to the same height before the volume is read can be appreciated by noting that raising or lowering the leveling bottle will give almost any desired volume in the measuring burette.

Another precaution that should be taken before the volume is read is to allow the water to drain down the sides of the burette for the same length of time. Under ordinary conditions half of a minute is enough. The inside of the measuring burette should be kept clean, in order that the water may drain well.

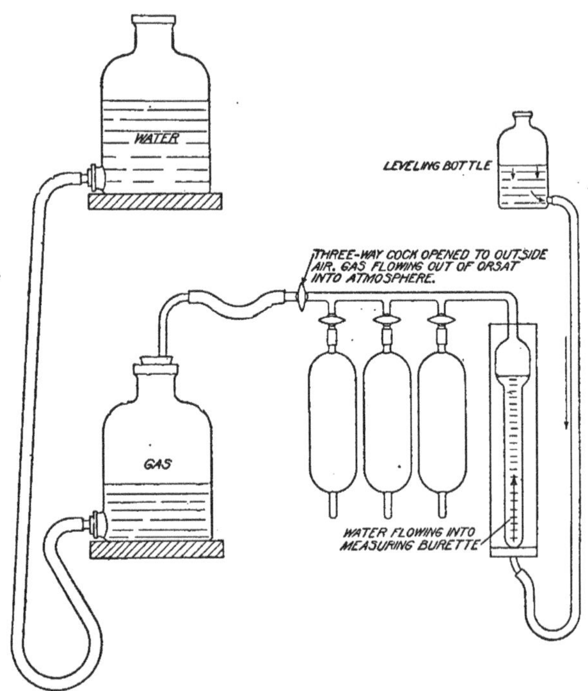

FIGURE 18.—Relative position of leveling bottle and the Orsat apparatus when gas is being expelled from the apparatus. The three-way cock is opened to the atmosphere.

TESTING FOR LEAKS.

Careful attention should be given joints on the Orsat apparatus. In making connections the ends of the glass tubing should be brought close together. The rubber joints must be perfectly tight. To test for leaks the measuring burette is about half filled with air and all the stopcocks are closed. The leveling bottle is then lowered about 2 or 3 feet, the air inside the burette being thus placed under a reduced pressure. If the water in the burette falls to a certain point and then remains stationary, there is no leak. If it continues to fall slowly, there is a leak, which must be stopped before the analysis goes further.

PREPARING ORSAT APPARATUS FOR ANALYSIS.

Before an analysis is begun each solution must be brought to the mark on the stem of the pipette and the corresponding valve in the header closed. Any gas that has been drawn into the measuring burette is expelled and the water is forced into the small-bore neck of the burette. The apparatus is then ready for use in an analysis.

PROCEDURE IN MAKING THE ANALYSIS.

CLEANING THE APPARATUS.

To take gas into the apparatus the end of the header is connected to the gas supply, such as a collecting bottle. The gas is drawn in by lowering the leveling bottle, with the three-way cock opened to the gas supply, as shown in figure 17. It is forced out by raising the leveling bottle with the three-way cock opened to the outside air, as shown in figure 18. Before a part for analysis is retained three or four burettes of gas are drawn into and forced out of the apparatus. The apparatus is thus cleaned of all the residue from the previous analysis, or of air, and a true sample of the flue gas to be analyzed can be obtained.

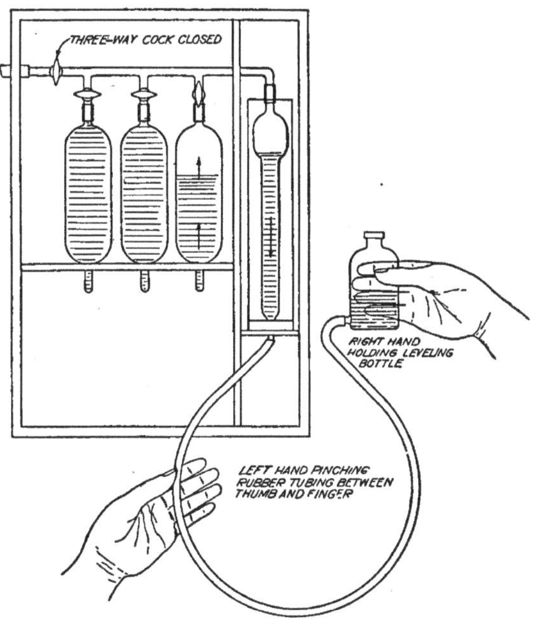

FIGURE 19.—Manipulation of Orsat apparatus when gas is being drawn out of absorption pipette. Left hand throttles the flow of water, thus controlling the flow of gas from the absorption pipette; right hand holds the leveling bottle.

MEASURING THE SAMPLE.

After three or four burettes of gas have been rejected, a burette of gas is drawn into the apparatus and the three-way cock is closed. The gas is placed under a slight pressure by raising the leveling bottle until the water is above the zero mark. The rubber tubing is then pressed between the thumb and first finger of the left hand and the leveling bottle is placed below the lower end of the measuring burette. The pressure exerted with the left hand on the rubber tubing is slowly released, and the water allowed to fall to the zero mark. It is held at this point by pinching the rubber tubing, while the three-way cock is opened to the outside air for a few seconds and closed. When the pressure on the rubber tubing is released and the water in the leveling bottle is held at the same level as the water in the burette, the level line should pass through the zero mark. If the line of common level does not pass through zero mark, the manipulation just described must be repeated.

CARBON DIOXIDE DETERMINATION.

When the volume of gas has been adjusted so that the water level passes through zero, the gas is forced into the pipette, containing potassium hydroxide solution. This is done in the manner indicated in figure 19, the leveling bottle being raised above the Orsat apparatus so that the water flows into the burette.

The rubber tubing is pressed between the thumb and first finger of the left hand, the leveling bottle placed on top of the Orsat apparatus, and the valve leading to the first pipette is opened. The pressure of the fingers on the rubber tube is released, allowing the water to run from the leveling bottle into the burette and forcing the gas into the pipette. When the water in the burette reaches the small-bore neck the leveling bottle is lowered and the gas is drawn back into the burette, care being taken not to draw the solution into the valve. The flow of gas can be nicely regulated with the pressure on the rubber tubing as described in connection with figure 19. This operation is repeated three times, the gas being slowly forced into and drawn out of the pipette.

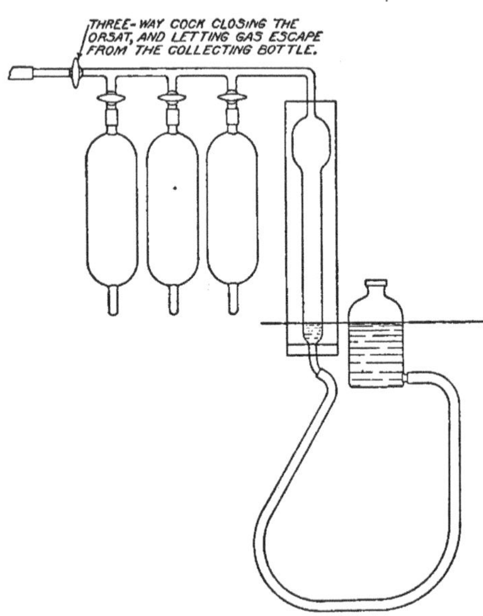

FIGURE 20.—Position of leveling bottle when gas volume is read. Water in measuring burette and leveling bottle at the same level.

After the gas has been drawn out of the pipette the last time the solution is brought to the mark on the neck and the stopcock is closed. The volume is measured by bringing the water in the leveling bottle to the water level in the measuring burette, as shown in figure 20. The gas is then forced again into the pipette and drawn back into the burette and the volume measured once more. If the two readings are the same, all of the carbon dioxide has been absorbed; if not, the operation is repeated until two readings are the same. The reduction in the volume of gas is due to the absorption of the carbon dioxide by the solution and is the measure of the amount of this constituent in the gas. It is well to remember that when the gas is drawn out of the pipette the solution should always be brought to the same mark on the stem and the valve closed before the volume is measured.

The solution can be brought to the mark with a nicety by throttling the water in the rubber tubing with the fingers of the left hand. In fact the transferring of the gas from the burette to the pipettes and back again can be done a great deal easier if the left hand is held constantly on the tubing, ready to throttle the flow of water whenever an exact adjustment of the levels is required. The manipulation is that indicated in figure 19. Care should be taken not to run water into the header, and particularly not to draw solution into the stopcocks.

If the glass tubes in the pipettes are not properly arranged, gas may pass through them into the other arm of the absorption pipette and be lost. The tubes should be watched when the gas is being passed into the pipette, and all gas loss avoided.

OXYGEN DETERMINATION.

After all the carbon dioxide has been taken out, the remaining gas is forced into the second pipette containing an alkaline solution of pyrogallic acid, which absorbs the oxygen. The gas is moved back and forth as in the absorption of carbon dioxide. Oxygen is absorbed much slower than carbon dioxide; therefore the gas must be passed into and out of the pipette a greater number of times. The absorption should be continued until two successive measurements are the same. The contraction in volume is the amount of oxygen in the gas.

CARBON MONOXIDE DETERMINATION.

Carbon monoxide is absorbed in an ammoniacal solution of cuprous chloride. After all the carbon dioxide and oxygen have been removed the carbon monoxide in the remainder of the gas is absorbed in the third pipette in the same manner as in the preceding determinations. Accurate determination of the carbon monoxide is difficult. Cuprous chloride solution, after being used for some time, will not remove all the carbon monoxide. Moreover, if gas containing no carbon monoxide is passed into a solution that has absorbed carbon monoxide previously some of it will be given off, and an increase in volume will result. To obtain accurate results the solution must be renewed frequently.

EXAMPLES OF ANALYSIS.

Analysis is started with the measuring burette filled with gas to the zero point. The gas is forced into and drawn out of the first pipette three times. The volume measures, say, 5.2. The gas is forced into and drawn out of the pipette once more. The volume is the same as before; therefore all the carbon dioxide has been absorbed. The contraction, or the carbon dioxide absorbed, is 5.2. The remaining gas is forced into and drawn out of the second pipette five times. The volume is found to be, say, 19.4. The gas is forced into and drawn out of the pipette twice more, and the volume is found to be, say, 19.6, showing that all the oxygen had not previously been absorbed. If, after two more passages, the volume remains 19.6, all the oxygen has been removed. The contraction is 19.6 minus 5.2, or 14.4, which is the amount of oxygen in the gas. The remaining gas is forced into and drawn out of the third pipette three times. The volume is measured and no contraction is observed; therefore no carbon monoxide is present.

An analysis is recorded in percentage by volume. As mentioned previously, the measuring burette of an Orsat apparatus is graduated to 100 units, the units being further graduated to fifths. If the sample fills the burette to the zero mark, the contraction can be read in terms of percentage by volume. If the sample of gas does not fill the burette to the zero mark, the contraction divided by the volume of gas taken for analysis and multiplied by 100 equals the percentage by volume. An example will illustrate: The volume of gas taken for analysis is 49.0 units. After carbon dioxide has been absorbed the volume is 44.2 units. The contraction is $49.0-44.2=4.8$ units. $\frac{4.8}{49.0}\times 100=9.8$ per cent of carbon dioxide. After the oxygen has been absorbed the volume is 39.8 units. The contraction is $44.2-39.8=4.4$ units. $\frac{4.4}{49.0}\times 100=9.0$ per cent of oxygen. It is much better, if possible, to fill the measuring burette to zero point.

PRECAUTIONS TO BE TAKEN TO OBTAIN ACCURATE RESULTS.

Accurate results can be obtained only by close attention to details. The apparatus must be perfectly tight. A test for leaks should be made each day before an analysis is begun. The constituents must be absorbed in the proper order—first, carbon dioxide; second, oxygen; and third, carbon monoxide. The solutions for oxygen and carbon monoxide absorb carbon dioxide and the solution for carbon monoxide absorbs oxygen also; therefore, each constituent must be removed completely before absorption of the next is begun. The solutions should be replaced by fresh ones when the absorption becomes slow. All measurements of volume must be made under the same pressure. Changes in temperature during an analysis should be avoided.

PREPARATION OF SOLUTIONS.

The absorbing agents employed are solutions that are prepared by dissolving a weighed quantity of chemicals in a measured volume of water. If no balance and graduate are available with which the materials can be weighed and measured, the quantities can be approximated in cubic inches with fair accuracy. However, it is advisable to provide an 8-ounce glass graduate, which will be found useful in measuring liquids. It can be bought from any chemical supply house for about 40 cents.

POTASSIUM HYDROXIDE SOLUTION FOR ABSORBING CARBON DIOXIDE.

For the preparation of the solution for absorption of carbon dioxide potassium hydroxide not purified by alcohol should be used. It can be bought in sticks about five-sixteenths of an inch in diameter. A stick 5 inches long weighs approximately one-half ounce. Three hundred and thirty grams of potassium hydroxide, about 14 ounces, or twenty-eight 5-inch sticks, are dissolved in 1,000 cubic centimeters, about 34 fluid ounces or 61 cubic inches, of clear water.

The solution can be made in a bottle or in an open vessel. If a bottle is used, the potassium hydroxide should be added to the water slowly to prevent breaking the bottle by the heat produced. If an open vessel is used, the water can be poured into the potassium hydroxide at once. The solution should be cooled before it is transferred to the bottle. If a precipitate forms, as is usually the case, it should be allowed to settle and the clear liquid poured off without disturbing the sediment.

Potassium hydroxide, thus prepared, can be kept in a rubber-stoppered bottle and used as a stock solution for carbon dioxide determinations and for making alkaline pyrogallic acid solution. A pipette full can be used for about 150 determinations of carbon dioxide.

ALKALINE PYROGALLIC ACID SOLUTION FOR ABSORBING OXYGEN.

The alkaline pyrogallic solution for absorbing oxygen is prepared by dissolving in the graduate or other glass vessel 10 grams (about one-third of an ounce or 2½ cubic inches) of pyrogallic acid in 25 cubic centimeters (about 1 fluid ounce or 2 cubic inches) of water. This is poured into the second pipette, and potassium hydroxide solution from the stock bottle is added until both arms of the pipette are a little more than half filled. The two solutions should be mixed in the pipette and the seal attached immediately to prevent absorption of the oxygen from the air. The mixing is done by raising and lowering the leveling bottle a few times and running gas into and out of the pipette.

Alkaline pyrogallic acid absorbs oxygen rapidly at first, but as oxygen is absorbed the speed of absorption decreases. The best results are obtained by renewing the solution whenever the absorption of oxygen becomes slow. One filling of the pipette can be used for about 40 determinations of oxygen in flue gas.

AMMONIACAL CUPROUS CHLORIDE SOLUTION FOR ABSORBING CARBON MONOXIDE.

The chemicals required for the preparation of a solution for the absorption of carbon monoxide are cuprous chloride, ammonium chloride, both small crystals, and ammonium hydroxide (ammonia water). The ammonia water should be clear and have a specific gravity of 0.91. The article sold in stores as household ammonia should not be used. It is best to purchase all the chemicals required from a chemical supply house.

A stock solution can be prepared best in a 2-quart bottle, which should be thoroughly cleaned before use. To 750 cubic centimeters (about 25 fluid ounces or 46 cubic inches) of water in the bottle is added 250 grams (about 9 ounces or 30 cubic inches) of ammonium chloride. The mixture is shaken until the salt crystals are dissolved. The ammonium salt dissolves readily and only a little shaking is necessary. To this solution is added 200 grams (about 7 ounces or 12 cubic inches) of cuprous chloride. The cuprous chloride dissolves slowly and the mixture should be shaken from time to time to aid solution.

Often the cuprous chloride does not all dissolve, and some of it settles on the bottom of the bottle. When the solution is being used, only the liquid should be poured off, without disturbing the sediment to any extent. A few pieces of clean scrap copper or copper wire placed in the bottle will help to keep the cuprous chloride reduced. If the bottle is tightly stoppered the solution can be stored indefinitely.

In preparing the stock solution for use enough is poured into the graduate or other glass vessel to fill one arm of the pipette. To this measured quantity is added slowly ammonium hydroxide (ammonia water). When the ammonia is poured in, a white flaky precipitate at first forms in the solution but disappears as more ammonia is added, and the solution becomes dark blue. The mixture should be stirred while the ammonia is added, and the amount of ammonia should be just enough to dissolve the precipitate completely.

Approximately one volume of ammonia is required for three volumes of the stock solution. If too much ammonia is added the ammonia vapor will increase the volume of the gas sample after contact with the solution. When all the precipitate has been dissolved the solution should be poured immediately into the pipette, as exposure to air weakens the solution. After the pipette has been filled according to the directions given in the section headed "Filling and Attaching Pipettes," any of the solution that may be left should be thrown away and not returned to the stock bottle.

A fresh solution of cuprous chloride, if given time enough, will remove all of the carbon monoxide from the gas mixture, but a solution that has absorbed considerable carbon monoxide will not remove all of this gas from a gas mixture and may even give off carbon monoxide. Therefore the operator must know the condition of the solution in order to be certain of results.

The length of time the solution can be used with good results will depend on the volume of carbon monoxide to be determined. For small amounts of carbon monoxide a solution that has not previously absorbed much carbon monoxide should always be used. Because of the small amount of carbon monoxide ordinarily present in flue gas and the difficulty of making an accurate determination, it is not advisable to make this determination regularly in boiler-room work, especially if the CO_2 content of the gases is low.

FILLING AND ATTACHING PIPETTES.

To clean and fill a pipette it should be removed from the stand. The required quantity of solution is poured in the open end of the pipette and the latter then fixed in position on the stand. The solution should fill both arms a little more than half full, and when drawn up to the mark on the stem of the pipette it should stand in the rear arm to a depth of about 1 inch. The front arm of the pipette is attached to the header by a piece of black rubber tubing about 1¼ inches long. To make the connection the rubber tubing is slipped entirely over the small end of the pipette, the glass tubes are brought together, and the rubber tubing is worked carefully upward until half of it is on the glass tube from the header and half remains on the pipette. Wetting the inside of the rubber causes it to slide easily on the glass.

The solutions are drawn up to the mark on the stem of the pipette by lowering the leveling bottle before beginning an analysis. The seal should be attached to the pipettes as soon as they are filled in order to protect the solutions from the air. When the joints on the Orsat apparatus must be broken it is best to cut the rubber tubing, because after a short time the rubber sticks to the glass and does not slide easily.

Care should be taken not to get the solutions on the hands and clothes, as they are strongly alkaline and will injure the clothes. When handling potassium hydroxide sticks cotton gloves or a piece of cloth can be used to protect the hands.

CARE OF ORSAT APPARATUS.

The operator will save time and expense and prevent many troublesome difficulties by taking good care of the Orsat apparatus.

If the ground-glass surfaces of stop cocks are allowed to stand without cleaning, they will become cemented together by alkaline solutions, and the cocks can not be operated. The only remedy is to keep the stopcocks free from alkali and lubricated with a thin film of vaseline. If too much vaseline is used, the openings in the cocks and capillary tubes become stopped with the excess. A properly lubricated stopcock has the appearance of a single piece of thick glass.

If a solution is accidently drawn into a stopcock, the cock should be removed at once and the surfaces wiped clean with a cloth or piece of soft paper, and lubricated with a thin film of vaseline. If necessary the header should also be removed and washed free from alkali.

The water in the burette and leveling bottle should be saturated with flue gas and should be changed as often as it becomes dirty. If the water becomes alkaline by solution being drawn into the header and washed into the burette it should be changed at once. If this is not done carbon dioxide will be absorbed by the alkaline water and the percentage of CO_2 indicated by the analysis will be low.

The joints made with rubber tubing should be examined and the apparatus tested for leaks before work is started. This is especially necessary when the Orsat apparatus is not used frequently.

The apparatus should be kept clean. If the walls of the burette and water jacket are dirty or dirty water is used, accurate measurements can not be made.

Skill in the use of the Orsat apparatus requires practice. However, an inexperienced but careful operator, after making about a dozen analyses can obtain results that are accurate enough for all ordinary requirements of boiler-room work.

USE OF GAS ANALYSIS IN REDUCING BOILER-ROOM LOSSES.

If gas analysis is intelligently applied to boiler furnaces it should prove itself an effective means of reducing boiler-room losses, for it furnishes information as to the magnitude and the cause of large chimney losses. An analysis for CO shows the approximate losses due to incomplete combustion. When efforts are made to reduce these losses, gas analysis shows how effective the applied remedies are.

Figure 21 is a graphic representation of the power-plant losses in an average industrial plant of 1,000 to 2,000 horsepower. The process of power generation, its transmission and utilization, is shown as a stream of heat starting with the coal fired under the boiler and ending in the power utilized in a cotton mill. The losses are shown as streams branching off to one side from the main stream. The first five side streams represent the boiler-room losses, and together amount to 43 per cent of the heat in the coal fired. The second loss, which is the heat carried away by the dry chimney gases, is the largest of boiler-room losses. It amounts to 26 per cent of the total heat in coal fired and is much larger than all the other boiler-room losses put together. It is this loss that, by the intelligent interpretation of gas analysis, can be reduced to 15 or even 10 per cent without increasing appreciably the loss from incomplete combustion.

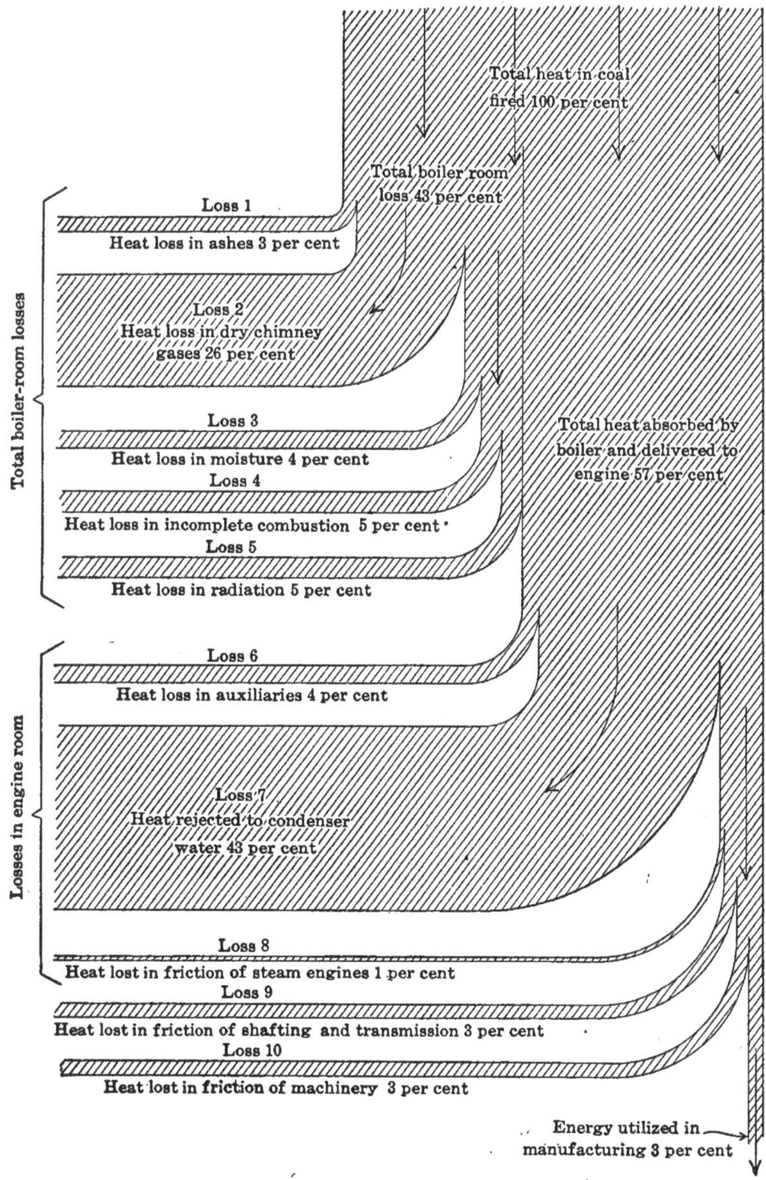

FIGURE 21.—Graphic representation of losses in an average industrial power plant having 1,000 to 2,000 horsepower capacity. The branches turning to the left from the main stream represent the losses.

For the complete combustion of 1 pound of coal in the ordinary boiler furnace about 15 pounds of air is necessary. This air is taken into the furnace at atmospheric temperature and leaves the boiler at a temperature of about 500° F. higher. It carries along with it all the heat that has been absorbed in raising its temperature 500° F. Each pound of air or of the products of combustion absorbs approximately one-fourth of a heat unit per degree of temperature rise. Therefore, with a temperature elevation of 500° F. each pound absorbs $\frac{500}{4}$, or 125, heat units.

If 15 pounds of air is used to burn 1 pound of coal the products resulting from the combustion weigh 16 pounds. The quantity of heat carried out with the chimney gases for each pound of coal burned then is 16 times 125, or 2,000 , heat units. If 1 pound of coal as fired contains 14,000 heat units, the heat carried out with the chimney gases constitutes $\frac{2000}{14000}$, or 14.3 per cent, of the total heat in the coal.

If instead of using 15 pounds of air to burn 1 pound of coal, the fireman uses 31 pounds of air, the products of combustion weigh 32 pounds, and the heat loss in the dry chimney gases is 32 times 125, or 4,000, heat units, or 28.6 per cent of the total heat in the coal fired. Thus it can be shown that the chimney losses increase almost directly with the amount of air used for the combustion of coal.

Besides increasing the chimney losses, the use of a large excess of air reduces appreciably the horsepower that can be developed with a given boiler installation. This feature is an important one in case the boiler plant is heavily loaded.

It is apparent then, that if the chimney losses are to be low, and the horsepower developed by the boiler high, the weight of air used in the combustion of coal must be kept as low as completeness of combustion will permit; 15 pounds of air to 1 pound of coal, if properly introduced into the furnace and with careful firing gives practically complete combustion in most boiler furnaces, but an engineer should find out how much air his fireman is using and whether he is getting complete combustion. It is just this information that the analysis of flue gases furnishes.

Combustion of coal is a chemical combination of its carbon and hydrogen with the oxygen of air, the products of combustion are carbon dioxide (CO_2) and water vapor (H_2O) respectively. Air consists approximately of 20 per cent of oxygen and 80 per cent of nitrogen by volume. If all the oxygen that enters the furnace were used in combustion, the analysis of the products would show about 81.5 per cent of nitrogen and 18.15 per cent of carbon dioxide. The percentage of carbon dioxide would not be as high as the percentage of oxygen in the air admitted into the furnace because some of the oxygen combines with the free hydrogen of coal and forms water vapor which is condensed and therefore does not appear in the analysis of gases.

If only one-half of the air entering the furnace is used in the combustion of coal the analysis of the products shows about 9 per cent of carbon dioxide and 10 per cent of free oxygen—that is, about half of the oxygen appears in the form of CO_2 and half as O_2. The CO_2 is not exactly equal to the O_2, because a small part of the oxygen used in combustion combines with the hydrogen of the coal and is condensed as water. Thus the CO_2 content of the flue gases shows what proportion of the air entering the furnace is actually used in the process of combustion and the oxygen content shows what proportion is in excess of the amount actually used.

CALCULATION OF THE WEIGHT OF GASES.

From the analysis the weight of the gases (W) per pound of coal burned can be computed from the following formula:

$$(1)\ W=\frac{11\,CO_2+8O_2+7(CO+N_2)}{3(CO_2+CO)}\times\frac{\text{per cent of carbon in coal}}{100}$$

Or, as $CO_2+O_2+CO+N_2=100$ always, with the simple Orsat analysis under consideration,

$$(2)\ W=\frac{(4\times CO_2+O_2+700)}{3(CO_2+CO)}\times\frac{\text{per cent of carbon in coal}}{100}$$

In applying equations *1* and *2* the percentages of CO_2, O_2, CO and N_2 in a given flue gas are to be substituted for the chemical symbols.

The following examples show the use of these equations:

Example A.

Suppose that the analysis of a given flue gas shows the following results:

	Per cent.
CO_2	7.0
O_2	12.0
CO	.2
N_2	80.8

Suppose the carbon content of the coal burned to be 85 per cent.

Substitution of the analytical results in equation *1* gives:

$$\frac{11\times7.0+8\times12.0+7(0.2+80.8)}{3(7.0+0.2)}\times\frac{85}{100}=\frac{740}{21.6}\times0.85=29.08$$ pounds of gases per pound of coal.

Substitution of the analytical results in equation *2* gives:

$$4\times7.0+12.0+700\times\frac{85}{100}=\frac{740}{21.6}\times0.85=29.08$$ pounds of gases per pound of coal.

Both equations give the same result; however, equation *2* is the simpler one to use.

EXAMPLE B.

Suppose that the analysis of a given flue gas shows the following results:

	Per cent.
CO_2	12.0
O_2	7.0
CO	0.4
N_2	80.6

Suppose the carbon content of the coal burned to be 85 per cent.
Substitution of the analytical values in equation *1* gives:

$$\frac{11\times12.0+8\times7.0+7(0.4+80.6)}{3(12.0+0.4)}\times\frac{85}{100}=\frac{755}{37.2}\times0.85=17.27$$ pounds of gases per pound of coal.

CALCULATION OF CHIMNEY LOSSES.

From the weight of gas per pound of coal the chimney losses can be computed by using the following equation:

(*3*) $W\times0.24\times(T-t)$=heat carried away with dry chimney gases expressed in British thermal units.

Where W=weight of gases in pounds per pound of coal as computed by equations *1* and *2*.
0.24=specific heat of gases.
T=temperature of gases leaving boiler, in °F.
t=temperature of air entering furnace, in °F.

If it is desirable to express this loss in percentage as given in figure 21, the value obtained by equation *3* is divided by the number of British thermal units in one pound of coal and multiplied by 100.

Thus in examples A and B if the temperature of air entering the furnace is 75° F., the temperature of the gases leaving the boiler 600° F., and the heat in the coal burned 13,500 British thermal units, the losses are:

$$(a)\quad \frac{29.08\times0.24\times(600-75)\times100}{13{,}500}=27.1 \text{ per cent.}$$

$$(b)\quad \frac{17.27\times0.24\times(600-75)\times100}{13{,}500}=16.1 \text{ per cent.}$$

In example A, with the CO_2 content in the flue gases 7.0 per cent, the loss is 27.1 per cent, and in example B, with 12 per cent of CO_2 in the flue gases, the loss is only 16.1 per cent of the heat in the coal, the difference being 11 per cent in favor of the furnace represented in example B having the high percentage of CO_2.

The two examples show that there is a definite relation between the percentage of CO_2 in the flue gases and the chimney losses represented in figure 21 by loss 2. This relation is shown in figure 22 in which the horizontal distances represent the percentages of CO_2 in flue gases and the vertical distances (the ordinates) represent the heat losses in dry chimney gases. Each of the four curves gives the heat losses for one temperature difference between the flue gases and the air entering the furnace, as indicated by the figures at the right of the curves.

The curves are computed for coal having 82 per cent of carbon and a heat value of 13,500 British thermal units; that is, a good grade of bituminous steam coal.

The following examples show how figure 22 can be used for computing dry chimney gas losses, when the percentage of CO_2 and the temperature of the flue gases and of the boiler-room air are determined.

EXAMPLE A.

CO_2 in flue gases =6.0 per cent.
Temperature of flue gases =600° F.
Temperature of boiler-room air=75° F.

The difference between the flue gas and the boiler-room temperature is 600°−75°=525° F. Therefore the second curve from the top, designated by the figures 525° F., should be used for this example. Take the vertical line starting from the point of 6.0 per cent of CO_2 on the scale at the bottom of the chart and follow it to the second curve from top; from the intersection point on this curve follow a horizontal line to the scale at the left, which gives the losses directly in percentage of the heat in the coal. In this case the losses in the dry chimney gases are found to be 31.5 per cent of the heat in the coal.

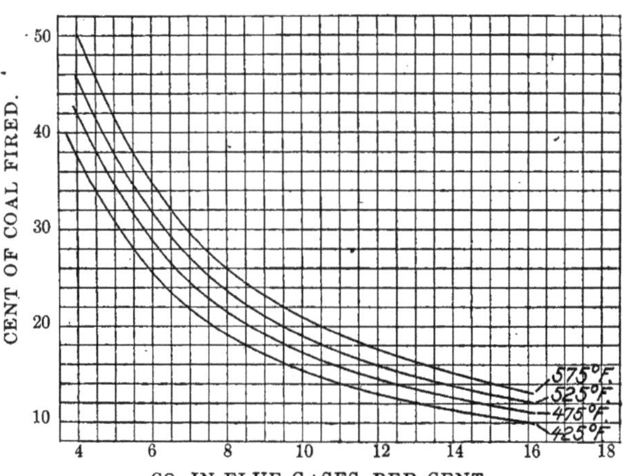

FIGURE 22.—Relation between the percentage of CO_2 in the flue gases and the heat lost in the dry chimney gases. Each of the curves gives this loss for one temperature difference between the flue gases and the air entering the furnace as indicated by the figures at the right of each curve.

EXAMPLE B.

CO_2 in flue gases =14.0 per cent.
Temperature of flue gases =650° F.
Temperature of boiler-room air=75° F.

The temperature difference between the flue gases and the boiler-room air is 575° F. Therefore the top curve should be used for this example.

Follow the vertical line of 14 per cent of CO_2 to the highest curve and from the intersection point a horizontal line to the scale at the left. The heat loss in the dry chimney gases is found to be 15 per cent.

The general characteristic of the curves is that as the CO_2 in the flue gases increases, the heat loss in the dry chimney gases decreases, at first very rapidly, but this decrease becomes small in the region of a high percentage of CO_2. For ordinary boiler furnaces probably it does not pay to run the CO_2 higher than 10 or 12 per cent. In many cases the leakage in the setting is such that the attainment of even 10 per cent is impossible.

CAUSES OF LOW CO_2 CONTENT.

The two principal causes of a low percentage of CO_2 are holes in the fire and leaky settings.

HOLES IN FIRES.

Technical Paper 80[a] describes the condition under which holes are formed in the fires, and points out the precautions that the fireman should use to prevent their formation. It suffices to repeat here that by firing small charges at short intervals of three to six minutes and by placing the fresh coal on the thin spots the formation of holes in the fuel bed can be prevented. It is not necessary to keep the fuel bed thicker than 5 or 6 inches. In fact, a 5-inch fuel bed if kept level may give higher percentages of CO_2 than a fuel bed 12 inches thick. The fireman is especially warned against heaping the coal on the front part of the grate and leaving the rear part uncovered.

LEAKY SETTINGS.

In the majority of instances leaky settings are the cause of a low percentage of CO_2 in the flue gases. This cause is harder to remove than the holes in the fires. Even a well-cared-for boiler setting allows a considerable volume of air to leak in. In the average plant probably not less than 30 per cent of the gases passing through the uptake leaks in through the setting. This feature is forcibly brought out by the gas analyses given in Tables 1 and 2 (pp. 51 and 52).

Table 1 gives the composition of the products of combustion as they pass through the uptake. Table 2 gives their composition as they enter a Heine boiler. The setting was seemingly in good condition; all the cracks as well as the joints between metal and brick wall were carefully packed with asbestos rope and the walls were painted with thick asphalt paint. In spite of all these precautionary measures considerable air entered the setting, as a comparison of the average values of the two tables shows. No serious objection can be made against the method of sampling the gases. Both sets of samples were collected simultaneously.

Similar difficulties are discussed in Bulletin 23.[b] All these data would seem to indicate that in plants where little attention is given to boiler settings the air leakage must be large indeed, and that it is important to keep boiler settings free from leaks as far as practicable if large chimney-gas losses are to be avoided.

In individual instances such losses can be approximately ascertained by taking simultaneous samples near the furnace, before the gases are diluted with the leakage, and in the uptake. It is advisable

[a] Kreisinger, Henry, Hand firing soft coal under power-plant boilers: Tech. Paper 80, Bureau of Mines, 1915, 83 pp.

[b] Breckinridge, L. P., Kreisinger, Henry, and Ray, W. T., Steaming tests of coal and related investigations, Sept. 1, 1904, to Dec. 31, 1908: Bull. 23, Bureau of Mines, 1912, pp. 289–293.

to take several sets of samples, the sampling tube being placed each time at a different point of the same cross section of the gas path. In estimating the leakage all sets of samples should be given consideration.

USE OF WATER-COOLED SAMPLER.

For the collection of a gas sample near the furnace it may be necessary to use a water-cooled sampler similar to the one shown in figure 23. Additional details of construction are shown in figure 27. Whenever possible this sampler should be used in a horizontal position. When used in this manner the trouble of the sampler becoming stopped with water condensation, soot, and tar will be greatly reduced.

The water inlet of this sampler must be connected to a cold-water supply of about 25-pounds pressure with a ½-inch garden hose, and

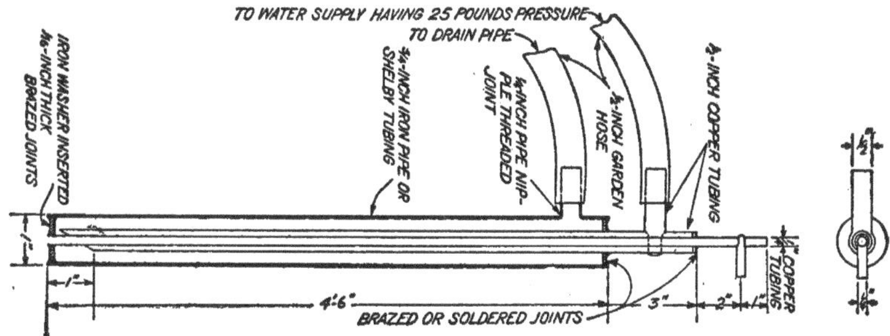

FIGURE 23.—Water-cooled gas sampler. Additional details are shown in figure 27.

the discharge from the sampler should be similarly connected to a drain. Both hose connections should be long enough to permit the sampler being inserted into or withdrawn from the furnace while the water is flowing through.

LOSSES FROM INCOMPLETE COMBUSTION.

In figure 21 the fourth side stream represents the total losses from incomplete combustion. By analyzing the flue gases for CO_2, O_2, and CO, the incomplete-combustion losses due to carbon burning to CO instead of to CO_2 can be determined. Such losses constitute perhaps one-half of the total incomplete-combustion losses. When 1 pound of carbon burns to CO_2, the combustion generates 14,500 British thermal units; when it burns to CO, only 4,500 heat units are generated. Thus 10,000 heat units are lost for every pound of carbon burning to CO. The gas analysis and the percentage of carbon in the coal furnish data for the computation of how much of the carbon in the coal fired burns to CO. With this quantity determined the heat loss due to this part of incomplete combustion can be obtained by multiplying by 10,000, which in round numbers is the heat value of 1 pound of carbon in the form of CO.

CALCULATION OF LOSSES DUE TO THE ESCAPE OF CO IN FLUE GASES.

The following are general formulas for the calculation of losses due to CO in flue gases:

(4) $$W=\frac{CO}{CO \text{ and } CO_2}\times\frac{C}{100}$$

where W=weight of carbon in pounds burned to CO for each pound of coal fired.
CO=percentage of CO in flue gases.
CO_2=percentage of CO_2 in flue gases.
C=percentage of carbon in coal.

(5) $$H=W\times 10{,}200$$

where H=heat loss in B. t. u. per pound of coal.
10,200=heat value in B. t. u. of 1 pound of carbon in the form of CO when burned to CO_2.

(6) $$L=\frac{H}{Q}$$

where L=the heat loss due to carbon burning to CO, expressed in percentage of heat in coal.
Q=heat value in B. t. u. per pound of coal.

The following examples show the method of calculating the heat losses due to CO in flue gases.

EXAMPLE A.

Suppose that the analysis of a given flue gas shows the following results:

	Per cent.
CO_2	7.0
O_2	11.2
CO	0.8

Suppose the carbon content of the coal burned to be 82 per cent. The heat value of 1 pound of the coal is 13,500 B. t. u.

Substitution of the values given by the analysis in equation *4* gives:

$W=\frac{0.8}{0.8+7.0}\times\frac{82}{100}=0.084$ pound of carbon burned to CO for each pound of coal fired.

$H=0.084\times 10{,}200=857$ B. t. u. lost per pound of coal burned.

$L=\frac{857\times 100}{13{,}500}=6.3$ per cent of the heat in coal fired is lost in CO.

EXAMPLE B.

Suppose that the analysis of a given flue gas gives the following results:

	Per cent.
CO_2	12.0
O_2	5.6
CO	1.4

Suppose the carbon content of the coal burned to be 82 per cent. The heat value of 1 pound of the coal is 13,500 B. t. u.

Substitution of the values given by the analysis in equation *4* gives:

$W=\frac{1.4}{1.4+12.0}\times\frac{82}{100}=0.085$ pounds of carbon burned to CO for each pound of coal fired.

$H=0.085\times 10200=867$ B. t. u. lost per pound of coal.

$L=\frac{867\times 100}{13{,}500}=6.4$ per cent of the heat in coal fired lost in CO.

The two specific examples show that the heat loss due to carbon burning to CO instead of CO_2 depends not only on the percentage of CO in flue gases but also on the CO_2 content. This fact is shown more clearly in figure 24 which shows the relation between the percentage of CO in the flue gases and the heat losses due to the escape of CO. Each of the curves gives this loss for one constant percentage of CO_2, as indicated by the figures under the curves. The curves are compiled for coal containing 82 per cent of carbon and having a heat value of 13,500 B. t. u. It is apparent from the position of these curves that the higher the percentage of CO_2 the lower is the heat loss for the same percentage of CO. Thus, with 1 per cent of CO and

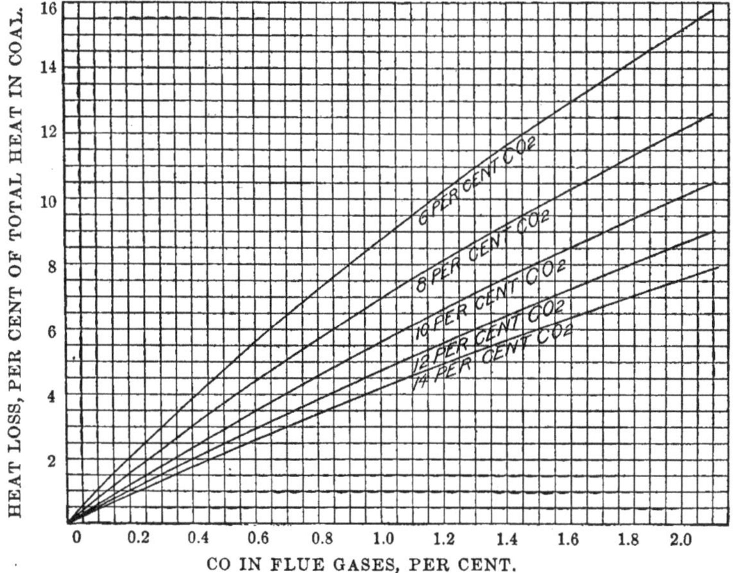

FIGURE 24.—Relation between the percentage of CO in the flue gases and the heat losses due to the escape of CO. Each curve gives the loss for one constant percentage of CO_2 as indicated under each curve.

6 per cent of CO_2, the loss is 8.8 per cent of the total heat in the coal, whereas if the percentage of CO_2 in the gases is 14, 1 per cent of CO causes only 4.2 per cent heat loss.

The curves of figure 24 can be used for computing approximately the incomplete combustion losses due to CO in flue gases. Thus, example A, on page 42, can be worked out as follows:

The vertical line starting from the point of 0.8 per cent at the bottom of the chart is followed to a point somewhat less than halfway between the curves of 8 and 6 per cent of CO_2. From this point a horizontal line is followed to the scale at the left which gives the heat losses for this example to be about 6.3 per cent, or the same as computed by the arithmetic.

Example B is worked out by means of the chart as follows:

From the point of 1.4 per cent of CO at the bottom of the chart follow the vertical line to the curve of 12 per cent of CO_2. From the intersection point of this curve follow the horizontal line to the scale at the left, which gives the losses to be about 6.4 per cent, or about the same as computed in the example.

Attention is called to the fact that CO constitutes only a part of the incomplete combustion losses. In burning soft coal when CO appears there are usually some hydrogen and gaseous hydrocarbons as well as tar vapors and soot escaping through the chimney along with CO. The tar vapors and soot form a large part of the visible smoke. Therefore the appearance of CO must be considered as only one indication of incomplete combustion. The total incomplete combustion may be twice as much or even more than that shown by the percentage of CO in the flue gases. In most plants the incomplete combustion losses are small compared to the heat carried away by the chimney gases; however, in some instances, these losses are great.

In the hand-fired furnaces CO appears in appreciable quantities one or two minutes after firing, as is shown in Table 5 (p. 57). Additional air supplied through the firing door immediately after firing tends to reduce the percentage of CO in the flue gases.

The determination of hydrogen and hydrocarbons requires more complicated apparatus than the Orsat and necessitates great skill on the part of the operator. Therefore it is not discussed in this report.

CAUSES OF INCOMPLETE COMBUSTION.

The chief causes of incomplete combustion are an insufficient supply of air admitted above the fuel bed and lack of proper mixing of this air with the combustible rising from the fuel bed. Perhaps in ordinary boiler furnaces the lack of proper mixing is more often the cause of incomplete combustion than the insufficient supply of air.

The amount of air introduced into the furnace, both through the grate and the firing door, can be approximately determined by the analysis of a gas sample taken in the furnace, or in some other place within the setting before the gases are diluted by the air leakage through the setting. Such samples should be collected with a water-cooled sampler similar to that shown in figure 23.

Several samples must be collected and analyzed, and the conditions of fire under which the samples are collected must be carefully observed and considered before the true condition of the combustion can be ascertained. Air leaking into the setting beyond the furnace seldom helps the combustion if combustible material is contained in the gases, because the temperature is usually too low for the ignition and burning of the combustible material. Therefore the analysis of a gas sample collected in the uptake may show both incomplete combustion and a large excess of air.. In the average plant, if reasonable care is taken of the boiler setting, 10 to 12 per cent of CO_2 in the flue gases gives about the best results as regards excess of air and incompleteness of combustion.

In some plants the CO_2 may reach as high as 14 per cent with only traces of incomplete combustion, whereas in other plants even 8 per cent of CO_2 may be accompanied with a considerable loss from incomplete combustion.

DESIRABILITY OF ANALYZING FLUE GASES.

The question may be asked, Is it desirable in all steam plants to analyze flue gases? In general, it can be said that it is desirable. Whether the gases should be sampled and analyzed continuously or periodically depends perhaps on the size of the plant and the nature of the load. In some of the large plants generating power 24 hours a day gases are sampled and analyzed continuously, and the fireman is paid a bonus according to how good a combustion he is able to obtain, the quality of the combustion being determined by analysis of the flue gases. The gas samples are analyzed either with an Orsat apparatus or with an automatic CO_2 recorder. This plan should be tried wherever it appears at all feasible.

In some plants, particularly those of smaller capacity and those operating with full load only during the day hours, it may perhaps not be practicable to sample and analyze gases continuously. In such plants gas sampling and analyzing should be applied long enough to determine definitely the prevailing furnace conditions and the efficiency of the methods of firing. When such determination has been made the proper remedies to increase the efficiency of the steam plant can be applied. In hand-fired plants such remedies consist chiefly of getting the fireman into the habit of firing small charges at short intervals and keeping the fuel bed free from holes by placing the fresh coal on the thin spots. Small and frequent firing makes the amount of combustible rising from the fuel bed nearly uniform and proportional to the constant air supply, so that incomplete combustion is thus reduced to a minimum. By keeping the fuel bed free from holes a large excess of air in the furnace is avoided. Stopping air leakages into the setting further helps in reducing the excess of air. After the efficient firing has been well established and the boiler settings put into good condition, the analyzing of the gases may be stopped and used only at intervals to check the fireman and to find whether the settings remain in good condition.

EXPERIMENTAL RESULTS OF SAMPLING AND COLLECTING FLUE GASES.

The recommendation of the methods of sampling and collecting flue gases already described are based on experimental data. Some of these data have been accumulated during investigations of the combustion of coal; other data have been obtained from experiments made especially for the purpose of obtaining information on the subject of sampling and collecting flue or furnace gases. Inasmuch as these data show plainly what accuracy can be expected when

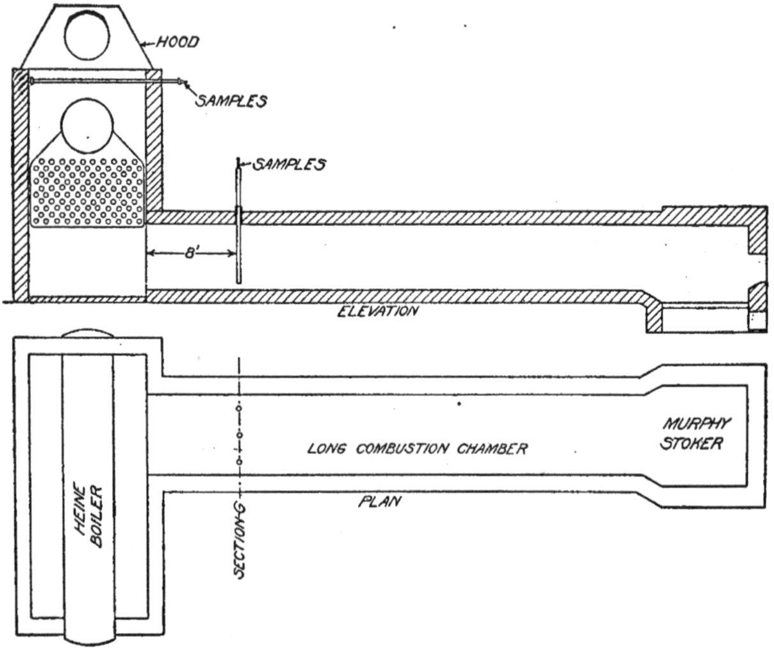

FIGURE 25.—Outline of experimental furnace with long combustion chamber connected to a Heine boiler. Section G in long combustion chamber is 8 feet from the entrance of gases into the Heine boiler.

flue gases are being sampled with various devices, the data are here presented and discussed.

The experiments, according to what they are intended to illustrate, are grouped into three classes, as follows:

(*a*) Experiments with different devices for sampling gases in water-tube and fire-tube boilers.

(*b*) Experiments showing the difficulty of collecting gases at a uniform rate over periods of three to five hours.

(*c*) Experiments on the absorption of CO_2 by water and brine used in collecting bottles.

EXPERIMENTS ON SAMPLING FLUE GASES.

Three series of tests were made to determine when the open-end single-tube sampler can be relied on to collect approximately the average sample, and when it is advisable to use the perforated tube sampler or some other more complex device. The first series of tests was made on a Heine boiler into which were discharged the hot products of combustion from a special experimental furnace. This furnace consisted essentially of a Murphy stoker, with a grate area of 25 square feet, and a combustion chamber 3 by 3 feet in cross section and about 45 feet long. The end of the combustion chamber was connected to the side of the boiler. The outline of the combined apparatus is shown in figure 25.

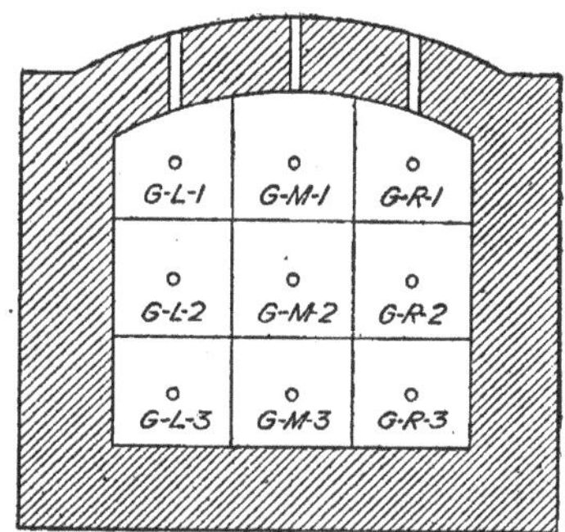

FIGURE 26.—Placing of samplers at section G of the long combustion chamber.

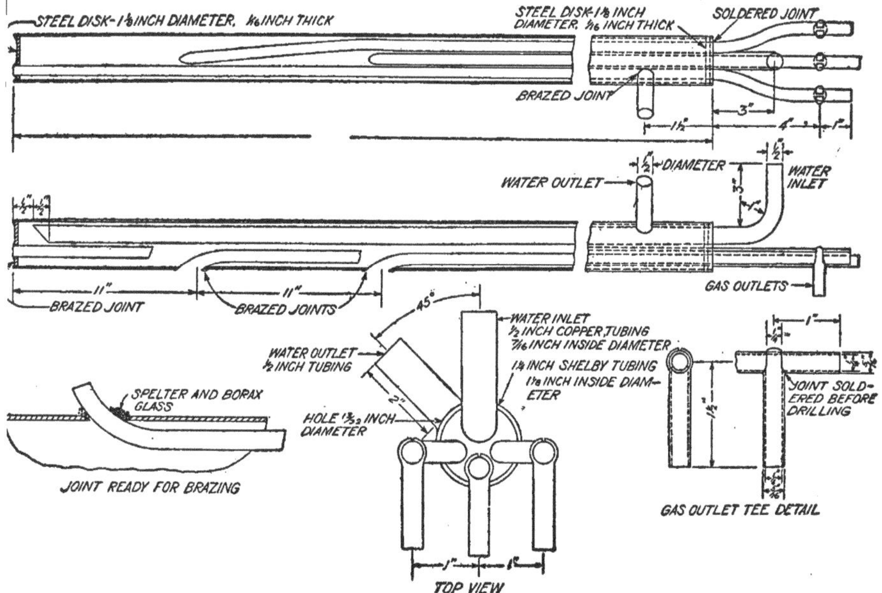

FIGURE 27.—Water-cooled gas sampler having three gas tubes in one water jacket. All joints that come in contact with flames are brazed; joints that remain outside of the furnace are soldered. When the sampler is being brazed or soldered, it should be held in such a position that the molten spelter or solder runs into the joint; this is important if good joints are to be obtained. When the joints to be brazed are being fitted, care should be taken to have the end of the copper tubing extend out far enough so that the spelter will not run down into it. After the joint has been brazed, the protruding end of the copper tube can be filed off. The side outlets on the gas tubes are first fitted as shown and then soldered by placing a piece of solder into the side outlet and heating the joints over a gas-torch flame, the side outlet being held up so that the molten solder runs all around the joint. The hole between the side outlet and the gas tube is drilled after the joint has been soldered. The gas sample is drawn through the side outlet, and the opening at the end of the gas tube is closed with a piece of rubber tubing and a glass plug and is used only when the tube is being cleaned after it has become stopped with soot.

The products of combustion were sampled at section G just before they entered the boiler and again in its uptake, where three different sampling devices were installed. At section G nine open-end sampling tubes were placed as shown in figure 26, so that each tube sampled gases from approximately 1 square foot of cross-sectional area.

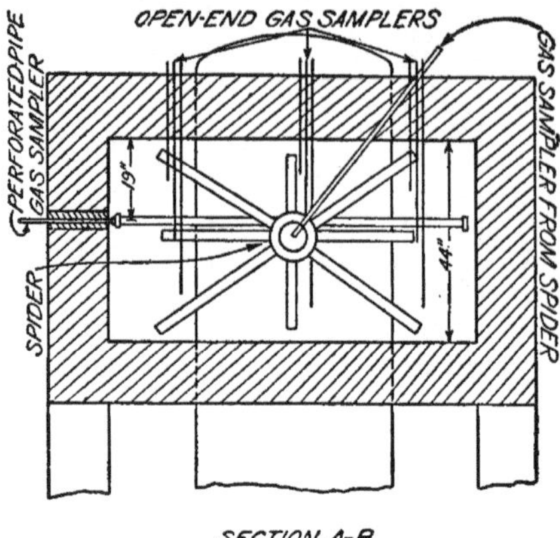

The sampling tubes were inserted through the holes in the arch. The samplers were water cooled, three tubes being inclosed in one water jacket as shown in figure 27. Through each of the sampling tubes the gas was collected into a separate gas holder.

In the uptake were placed nine open-end sampling tubes, one perforated sampler, and a special sampling device, which, on account of its shape, was called a "spider." The placing of these devices is shown in figures 28, 29, and 30.

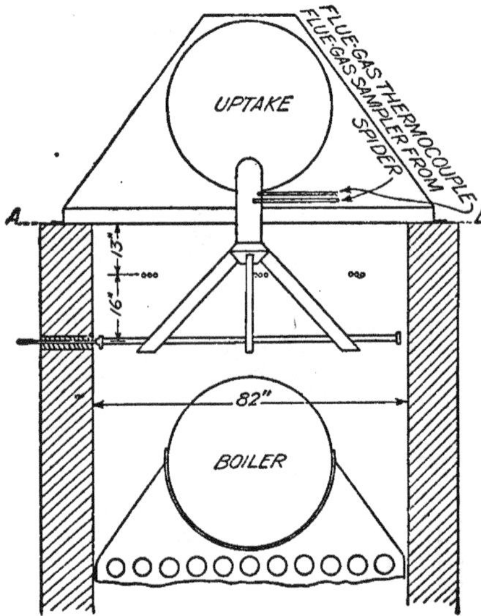

FIGURE 28.—General arrangement of three kinds of sampling devices in the uptake of a Heine boiler connected to long combustion chamber.

The open-end tubes consisted of standard ¼-inch pipe. They were inserted into the uptake through three holes in the rear wall.

The perforated-tube sampler was of the same design as that shown in figure 5. It consisted of a piece of 1¼-inch pipe 7 feet long, closed at both ends. The small tube extending to its center was standard ¼-inch pipe. The perforations were one-sixteenth of an inch in diameter and 3 inches apart. This sampler was inserted into the uptake through a hole in the side wall.

The "spider" was a device for obtaining the average sample and the average temperature of flue gases. It was made of No. 20 galvanized sheet iron. The details of its construction are shown in figure 31.

The gas entered the device through the long narrow opening in the bottom of the "legs." The streams of gas from the eight legs came together in the conical body and left through the 6-inch pipe bent into the breeching. A ¼-inch pipe was inserted into the center of the 6-inch pipe for drawing the sample. Immediately above this ¼-inch pipe was placed the junction of a thermocouple for measuring the temperature of the flue gases.

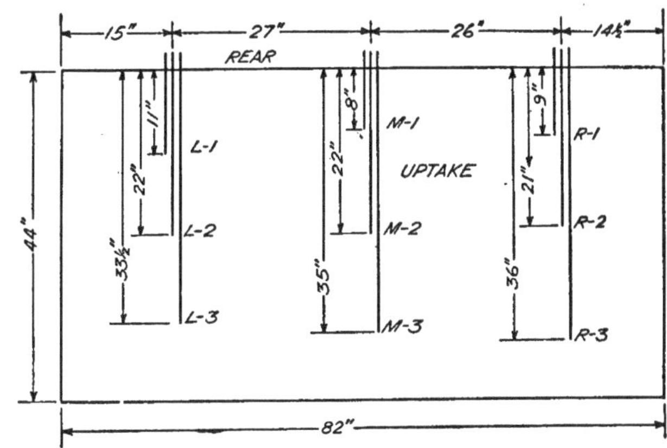

FIGURE 29.—Placing of nine open-end samplers in uptake of Heine boiler.

Each of the nine open-end sampling tubes, the perforated sampler, and the ¼-inch pipe taking gas out of the "spider" were connected to separate gas-collecting devices, one of which is shown in figure 32.

The collecting devices were connected with a 1¼-inch pipe to a strong steam ejector which caused a continuous stream of gas to flow through each of the sampling tubes. The gas streams were caused to bubble through the water in the wash bottle, so that if any of the sampling tubes became clogged the bubbling in the wash bottle ceased. Thus no sample could be missed without the operator knowing it. The gas was collected over mercury in glass holders of 125 c. c. capacity. The design of the gas holder is given in Plate I, which shows a table and equipment for collecting nine samples.

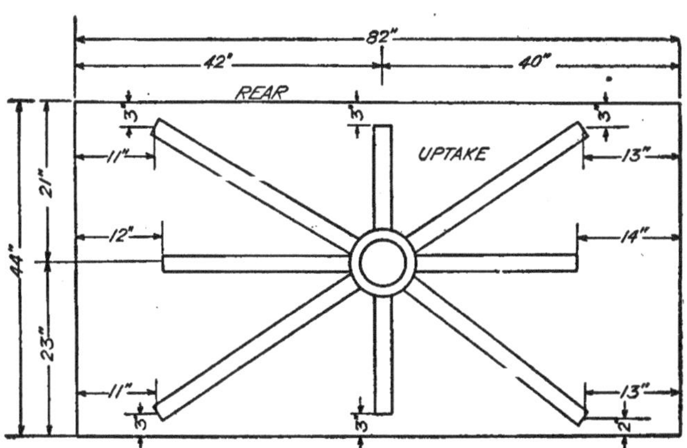

FIGURE 30.—Placing of the "spider" in uptake of Heine boiler.

The gas holders are mounted in wooden stands, each of which contains three gas holders. The figure shows one of the stands removed, so that the wash bottles and their connections may be seen more clearly.

All samples were collected simultaneously and the time of collecting them was about 20 minutes.

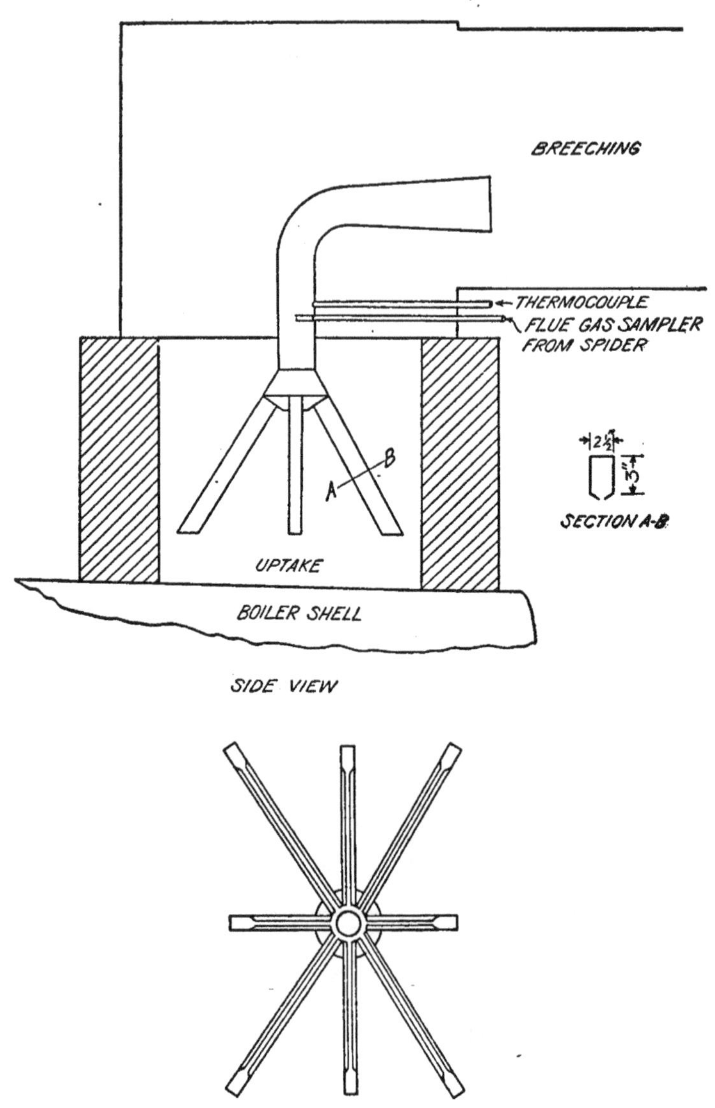

FIGURE 31.—Details of construction of "spider." Bottom view shows openings through which gases enter "spider." Side view shows placing of "spider" with respect to hood and breeching.

Tables 1 and 2 following show the composition of eight sets of simultaneous samples collected from the Heine boiler connected to the experimental long combustion chamber.

EQUIPMENT FOR COLLECTING NINE SAMPLES OF FLUE GASES.

The nine samples can be identified by reference to figure 29. The test number refers to the test with the experimental furnace.

TABLE 1.—*Results of analyses of samples taken in the uptake of a Heine boiler.*

Sample.[a]	Test 185.		Test 186.		Test 187.		Test 188.	
	CO_2	O_2	CO_2	O_2	CO_2	O_2	CO_2	O_2
L-1	5.8	14.2	7.9	11.7	8.9	10.4	8.7	10.8
L-2	5.9	13.9	7.9	11.8	8.9	10.3	9.1	10.2
L-3	5.5	14.5	8.3	11.4	9.3	9.7	9.8	9.6
M-1	5.7	14.3	7.6	12.0	8.8	10.7	8.7	10.7
M-2	5.7	14.4	7.9	11.7	8.7	10.5	8.8	10.6
M-3	5.5	14.5	8.1	8.2	9.1	9.9	8.8	10.4
R-1	6.7	13.0	9.8	9.5	9.8	9.6	9.7	8.9
R-2	6.0	14.0	8.7	10.8	10.1	9.0	9.9	9.4
R-3	5.0	15.0	7.4	12.3	8.3	11.3	8.3	11.3
Average	5.75	14.2	8.2	11.0	9.1	10.2	9.1	10.2
Maximum variation	5.0 to 6.7	....	7.4 to 9.8	....	8.3 to 10.1	....	8.3 to 9.9	....
(b)	5.6	14.6	8.1	11.4	8.9	10.3	8.8	10.5
(c)	5.5	14.1	8.4	11.2	9.2	10.0	9.4	10.0

Sample.[a]	Test 189.		Test 190.		Test 192.		Test 193.	
	CO_2	O_2	CO_2	O_2	CO_2	O_2	CO_2	O_2
L-1	9.0	10.7	8.4	11.0	11.0	8.0	11.3	6.8
L-2	10.1	9.4	9.2	10.2	10.6	8.7	9.7	9.3
L-3	10.9	8.1	10.6	8.6	9.7	9.8	10.1	8.8
M-1	8.8	9.1	9.0	10.3	9.3	9.6	9.9	9.0
M-2	9.9	9.7	9.3	9.8	9.9	9.2	10.7	8.2
M-3	9.4	9.9	9.2	10.1	9.8	9.8	10.5	8.5
R-1	11.1	7.2	10.5	8.8	9.2	10.4	9.5	9.1
R-2	11.1	7.9	10.2	8.8	10.3	9.7	11.0	7.8
R-3	9.1	10.5	8.2	11.3	11.5	8.8	11.9	6.7
Average	9.9	9.2	9.4	9.9	10.1	9.3	10.5	8.2
Maximum variation	8.8 to 11.1	....	8.2 to 10.6	....	9.2 to 11.5	....	9.5 to 11.9	....
(b)	....	....	9.1	10.2	9.5	9.5	10.5	8.3
(c)	10.1	9.4	9.5	9.9	....	....	....	....

[a] See figure 29, p. 49. [b] Sample taken with "spider." [c] Sample taken through 1¼-inch perforated pipe.

TABLE 2.—*Results of analyses of samples collected at section G of long combustion chamber.*

Sample.[a]	Test 185.			Test 186.			Test 187.			Test 188.		
	CO_2	O_2	CO	CO_2	O_2	CO	CO_2	O_2	CO	CO_2	O_2	CO
G-L-1	8.5	10.8	0.0	12.8	5.9	0.0	14.2	4.4	0.0	14.8	3.5	0.3
L-2	8.8	10.8	.0	12.9	6.2	.0	14.1	4.4	.0	15.0	3.5	.2
L-3	9.0	10.5	.0	13.3	5.6	.0	14.7	3.8	.0	14.9	3.6	.3
G-M-1	9.2	10.4	.0	13.0	5.6	.1	14.4	3.9	.0	14.8	3.6	.3
M-2	9.3	10.3	.0	13.1	5.6	.0	14.0	4.4	.0	14.1	4.3	.2
M-3	9.3	10.2	.1	13.2	5.7	.1	13.1	5.3	.0	14.9	3.6	.3
R-1	5.6	14.5	.0	11.5	7.5	.0	13.7	5.1	.0	13.1	5.4	.1
R-2	9.8	9.7	.0	13.2	5.7	.0	14.3	4.2	.0	15.0	3.4	.3
R-3	10.0	9.5	.0	13.4	5.6	.0	13.4	5.5	.0	14.8	3.5	.4
Average	8.8	10.7	.0	12.9	5.9	.0	14.0	4.6	.0	14.6	3.8	.3
Maximum variation	5.6 to 10.0	......	......	11.5 to 13.4	......	......	13.1 to 14.7	......	......	13.1 to 15.0	......	......

[a] See figure 26, p. 47.

TABLE 2.—*Results of analyses of samples collected at section G of long combustion chamber*—Continued.

Sample.[a]	Test 189.			Test 190.			Test 192.			Test 193.		
	CO_2	O_2	CO	CO_2	O_2	CO	CO_2	O_2	CO	CO_2	O_2	CO
G-L-1	14.6	4.1	0.0	14.3	4.3	0.0	13.5	4.9	0.0	13.7	4.7	0.0
L-2	14.6	3.8	.0	14.0	4.1	.0	13.9	4.3	.2	13.5	4.9	.C
L-3	14.6	4.0	.0	14.0	4.5	.0	13.3	5.3	.1	13.5	5.0	.1
G-M-1	14.7	3.6	.0	14.2	4.2	.0	14.3	3.9	.0	14.2	3.7	.0
M-2	[b].6	[b]20.2	.0	12.3	5.5	.0	15.2	2.4	.0	14.6	3.8	.0
M-3	14.7	4.0	.0	13.8	4.9	.0	11.3	7.2	.3	12.5	5.9	.0
R-1	12.5	6.5	.0	12.5	6.1	.0	11.1	7.8	.0	14.7	3.6	.1
R-2	14.8	3.5	.0	14.3	4.2	.0	14.9	3.4	.1	14.6	3.6	.0
R-3	15.0	3.4	.0	14.0	4.4	.0	15.1	3.3	.1	15.0	3.4	.0
Average	14.4	4.1	.0	13.7	4.7	.0	13.6	4.7	.1	14.0	4.3	.0
Maximum variation	12.5 to 15.0	......	.0	12.3 to 14.3	......	......	11.1 to 15.2	......	......	12.5 to 15.0	......	......

[a] See figure 26, p. 47.

[b] Not included in average; leakage in tube.

Table 1 shows that with the boiler equipment used the center sample, designated M-2, came as near to the average as the "spider" sample or the sample taken through the perforated tube. Therefore, with similar installations it would be safe to use the open-end tube placed in the center of the cross sections of the uptake.

As shown by the analysis of the samples collected at section G (Table 2), the gas is of a fairly uniform composition when it enters the boiler. The uniformity is undoubtedly the result of the long travel of the gas through the unusually long combustion space of the experimental furnace which gives the gases an opportunity to mix. If the gases enter the boiler uniformly mixed, it is to be expected that they leave it in a similarly uniform composition. It is only the air leakage into the boiler setting that could possibly destroy such uniformity of composition. Comparison of the analyses of gases collected in the uptake with those of the gases collected at section G will indicate that a considerable amount of air leaked into the setting. However, the mixing of the gases as they pass among the boiler tubes is such that the gases come out nearly uniformly diluted with air over the entire cross section of the uptake.

To determine whether the mixing of the gases as they pass through the boiler is sufficient to cause a uniform gas mixture in the uptake, a set of nine open-end sampling tubes and one perforated tube was placed in the uptake of a Heine boiler equipped with a Jones underfeed stoker. The stoker was placed directly under the boiler as in the standard commercial installation. The gases passed first under the boiler to the rear, then among the boiler tubes to the front, and then under the steam drum to the uptake in the rear of the boiler. The placing of the sampling tubes is shown in figure 33. The tubes were connected to the same gas-collecting apparatus as shown in figure 32 and Plate I.

Four sets of samples were collected in this installation. The time of collecting the samples was about 15 minutes. The analyses of the samples are given in Table 3. The table shows that the mixing of the gases as they pass among the boiler tubes is such that the

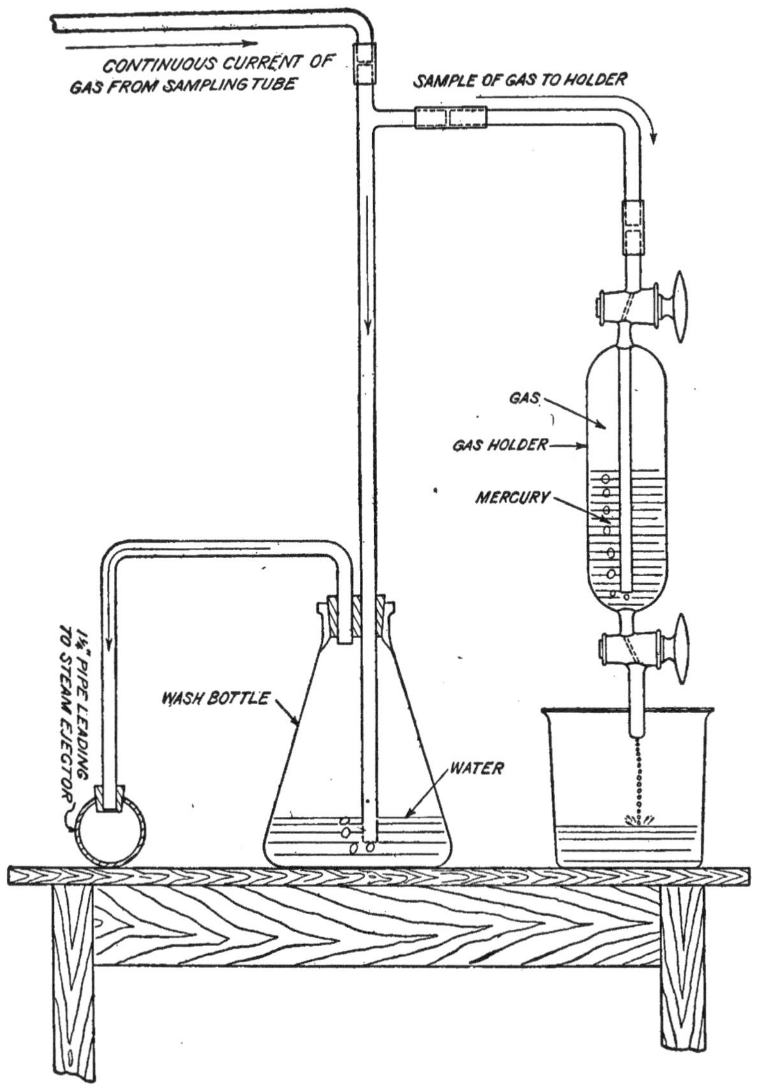

FIGURE 32.—Gas-collecting device in which gas is collected over mercury.

gas mixture in the uptake is fairly homogeneous. The composition of the sample collected with the central open-end tube was nearly the same as that of the sample collected with the perforated sampling tube, and both were close to the average of the nine samples.

TABLE 3.—*Results of analyses of samples taken in uptake of a Heine boiler equipped with Jones underfeed stoker.*

Sample.[a]	Set 1.				Set 2.			
	CO_2.	O_2.	CO.	CO_2 $+O_2$ $+\frac{1}{2}CO$.	CO_2.	O_2.	CO.	CO_2 $+O_2$ $+\frac{1}{2}CO$.
L-1	6.0	13.6	0.0	19.6	9.2	10.1	0.1	19.3
L-2	5.0	14.8	0.0	19.8	7.9	11.4	0.0	19.3
L-3	4.9	14.4	0.0	19.3	8.0	11.5	0.0	19.5
M-1	6.0	12.0	0.1	18.0	9.1	10.0	0.1	19.1
M-2	5.8	13.4	0.0	19.2	9.1	10.3	0.1	19.4
M-3	5.7	14.2	0.1	19.9	9.1	7.4	0.0	16.5
R-1	6.2	13.6	0.1	19.8	10.4	8.9	0.1	19.3
R-2	5.9	14.0	0.0	19.9	9.3	10.2	0.1	19.5
R-3	5.7	14.2	0.1	19.9	9.0	10.6	0.1	19.6
Average	5.7	13.8	0.0	19.5	9.1	10.1	0.1	19.2
Maximum variation	4.9 to 6.2				7.9 to 10.3			
[b]	5.9	14.0	0.1	19.9	9.1	10.3	0.1	19.4

Sample.[a]	Set 3.				Set 4.			
	CO_2.	O_2.	CO.	CO_2 $+O_2$ $+\frac{1}{2}CO$.	CO_2.	O_2.	CO.	CO_2 $+O_2$ $+\frac{1}{2}CO$.
L-1	10.5	8.8	0.3	19.4	11.8	7.2	0.2	19.1
L-2	8.6	10.6	0.4	19.4	10.8	8.3	0.1	19.1
L-3	9.3	10.0	0.1	19.3	11.1	8.0	0.1	19.1
M-1	10.8	8.3	0.0	19.1	11.0	6.3	0.0	17.3
M-2	10.2	9.0	0.0	19.2	11.8	7.2	0.2	19.1
M-3	10.2	6.8	0.0	17.0	12.0	5.4	0.0	17.4
R-1	11.0	7.9	0.1	18.9	12.6	6.2	0.0	18.8
R-2	10.4	8.9	0.0	19.3	11.9	7.0	0.1	18.9
R-3	10.0	9.2	0.1	19.2	11.8	7.1	0.1	18.9
Average	10.1	8.8	0.1	18.9	11.6	7.0	0.1	18.6
Maximum variation	8.6 to 11.0				10.8 to 12.6			
[b]	10.3	9.1	0.1	19.4	11.8	7.4	0.2	19.3

[a] See figure 33. [b] Sample taken through 1½-inch perforated pipe.

The preceding analyses of gas samples show that the gases in the uptake of a water-tube boiler form a fairly homogeneous mixture. This uniformity of mixture is ascribed to the fact that the gases mix freely as they pass among the tubes. In any fire-tube boiler no such mixing can occur; the gases in each tube stay separate as long as they stay in the tubes. Therefore, it would seem that the gases in the uptake of a horizontal tubular boiler would not be a uniform mixture, but that their composition would vary considerably in different parts of the uptake. To determine how much this variation is, a set of nine open-end sampling tubes was placed in the uptake of a 100-horsepower horizontal tubular boiler and several sets of simultaneous samples were collected and analyzed. The placing of the sampling tubes in the uptake is shown in figure 34.

The tubes were connected to the gas-collecting apparatus shown in figure 32. The sample from each tube was collected over mercury in a separate gas holder. The time of collecting the gas sample was about two minutes. Table 4 gives the analyses of the first four sets of samples. No sample was collected with a perforated-tube sampler. The analyses show that the composition varies over a wide range, and that the central sample as a rule contains a higher percentage of CO_2 than the average of the nine tubes.

Table 5 gives the analyses of four more sets of samples collected at various intervals after four equal and consecutive firings. Simultaneous samples were also collected with a perforated-tube sampler placed near the intakes of the nine open-end sampling tubes as shown in figure 34. The sample collected through the perforated sampler was nearer to the average than the sample taken with the center open-end tube; however, the difference was by no means small or negligible. The variation of the composition of the samples shows rather emphatically that it is very difficult, if not impossible, to collect a sample of flue gas that will represent the average composition of gases at a given cross section in the uptake of a horizontal return-tubular boiler.

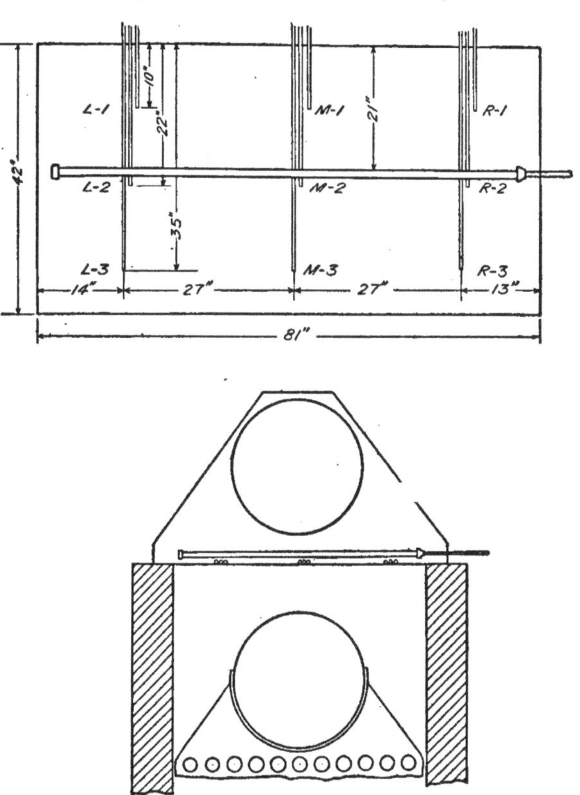

FIGURE 33.—Placing of nine open-end sampling tubes and a perforated-pipe sampler in a Heine boiler equipped with Jones stoker.

In general, it may be stated that flue or furnace gas samples as they are ordinarily taken do not represent the average composition of the entire stream of gas nearer than 0.5 per cent of CO_2, and in many instances the sample collected is 1 per cent or more of CO_2 higher or lower than the true average. It is therefore useless to make the analyses any closer than to the nearest 0.5 per cent of CO_2, particularly if no determination of the combustible gases is made.

TABLE 4.—*Results of analyses of samples collected in uptake of horizontal return-tubular boiler.*

Sample No.[a]	Set 1.			Set 2.				
	CO_2.	O_2.	$CO_2+O_2+\frac{1}{2}CO$.	CO_2.	O_2.	CO.	H_2.	$CO_2+O_2+\frac{1}{2}CO$.
1	10.1	8.6	18.7	12.1	7.3	....	....	19.4
2	12.4	6.5	18.9	14.8	4.0	....	....	18.8
3	12.2	6.8	19.0	15.3	3.3	....	....	18.6
4	12.9	5.3	18.2	16.0	2.6	0.8	0.0	19.0
5	13.2	5.8	19.0	15.0	3.4	1.0	0.2	18.9
6	9.0	10.3	19.3	13.4	5.4	0.5	0.4	19.0
7	8.6	10.7	19.3	11.8	7.4	....	....	19.2
8	8.3	11.2	19.5	12.7	6.5	....	....	19.2
9	9.0	10.3	19.3	13.4	4.4	....	....	17.8
Average	10.6	8.4	19.0	13.8	4.9	....	....	18.7
Maximum variation	8.3 to 13.2	....	....	11.8 to 16.0	....	....	....	....

Sample No.[a]	Set 3.						Set 4.			
	CO_2.	O_2.	CO.	H_2.	CH_4.	$CO_2+O_2+\frac{1}{2}CO$.	CO_2.	O_2.	CO.	$CO_2+O_2+\frac{1}{2}CO$.
1	9.0	7.3	5.5	2.3	0.4	19.1	12.8	7.2	0.1	20.0
2	10.7	4.9	6.6	3.1	.3	18.9	15.2	4.2	0.0	19.4
3	12.0	3.5	5.9	2.7	.4	18.5	15.7	4.6	0.0	20.3
4	12.5	2.7	5.9	2.6	.3	18.2	14.2	5.7	0.0	19.9
5	11.9	4.4	3.7	1.7	.1	18.2	12.7	7.0	0.0	19.7
6	10.3	7.2	2.8	1.3	.1	18.9	11.8	8.1	0.0	19.9
7	10.2	7.4	2.8	1.0	.3	19.0	10.7	9.8	0.0	20.5
8	10.1	7.4	2.7	.7	.4	18.9	10.8	9.2	0.0	20.0
9	11.5	6.2	1.9	.9	.3	18.7	10.2	9.3	0.1	19.5
Average	10.9	5.7	4.2	1.8	.3	18.7	12.7	7.2	....	19.9
Maximum variation	9.0 to 12.5	....	....	....	....	....	10.2 to 15.7	....	....	....

[a] See figure 34.

TABLE 5.—*Results of analyses of samples collected in uptake of horizontal return-tubular boiler at various intervals after firing.*

Sample No.[a]	Immediately after firing.						1½ minutes after firing.					
	CO_2.	O_2.	CO.	CH_4.	H_2.	$CO+CH_4+H_2$.	CO_2.	O_2.	CO.	CH_4.	H_2.	$CO+CH_4+H_2$.
1[b]	0.0	20.0	....	....	....	....	8.3	10.1	1.1	0.0	0.5	1.6
2	9.8	7.4	2.5	0.5	1.1	4.1	10.2	7.7	1.5	.0	1.2	2.7
3	9.8	7.4	2.5	.5	1.2	4.2	11.9	5.5	1.7	.2	.4	2.3
4	10.4	6.4	2.8	.7	1.3	4.8	12.3	5.1	1.7	.2	.5	2.4
5	9.8	7.2	2.7	.7	1.4	4.8	11.2	6.2	2.0	.3	.7	3.0
6	10.0	6.8	2.9	.8	1.8	5.5	11.2	6.0	2.2	.3	.9	3.4
7	8.6	8.4	2.9	.5	.9	4.3	8.9	8.8	1.9	.3	.9	3.1
8	7.4	10.0	2.2	.2	1.7	4.1	8.7	9.2	2.0	.4	.6	3.0
9	7.1	9.3	2.4	.4	1.4	4.2	8.4	9.2	2.0	.5	.4	2.9
Average	9.1	7.9	2.6	.5	1.4	4.5	10.1	7.5	1.8	.2	.7	2.7
Maximum variation	7.1 to 10.4	....	....	....	....	....	8.3 to 12.3	....	....	....	....	....
(c)	9.5	7.5	2.3	.4	1.5	4.2	10.4	6.7	1.5	.1	.5	2.1

[a] See figure 34. [b] Air leak. [c] Sample taken through 1-inch perforated pipe.

TABLE 5.—*Results of analyses of samples collected in uptake of horizontal return-tubular boiler at various intervals after firing*—Continued.

Sample.[a]	2½ minutes after firing.						3½ minutes after firing.					
	CO_2.	O_2.	CO.	CH_4.	H_2.	$CO+CH_4+H_2$.	CO_2.	O_2.	CO.	CH_4.	H_2.	$CO+CH_4+H_2$.
1[b]	9.2	9.6	0.4	0.0	0.0	0.4	9.5	8.6	0.8	0.0	0.2	1.0
2	11.7	6.5	.5	.0	.0	.5	11.2	6.4	.9	.0	.4	1.3
3	12.4	5.7	.2	.0	.0	.2	12.5	5.1	.9	.0	.4	1.3
4	12.8	5.5	.2	.0	.0	.2	12.4	4.6	1.5	......	......	......
5	11.7	5.9	.7	.0	.1	.8	11.7	5.8	1.0	.0	.1	1.1
6	11.0	6.4	.7	.0	.4	1.1	11.5	6.3	.8	.0	.1	.9
7	9.7	7.1	1.6	.1	.6	2.3	10.7	7.5	.4	.0	.1	.5
8	9.0	9.5	.6	.0	.2	.8	8.4	10.3	.2	.0	.1	.3
9	9.8	7.8	1.1	.0	.0	1.1	8.9	9.7	.3	.0	.1	.4
Average	10.8	7.1	.7	......	.1	.8	10.8	7.1	.8	.0	.2	1.0
Maximum variation	9.0 to 12.8	......	......	......	......	......	8.4 to 12.5	......	......	......	......	......
(c)	11.4	6.8	.2	.0	.1	.3	12.3	5.3	.9	.0	.4	1.3

[a] See figure 34. [b] Air leak. [c] Sample taken through 1-inch perforated pipe.

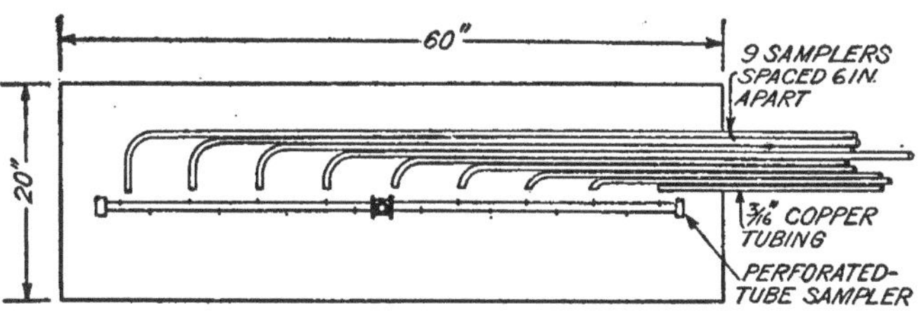

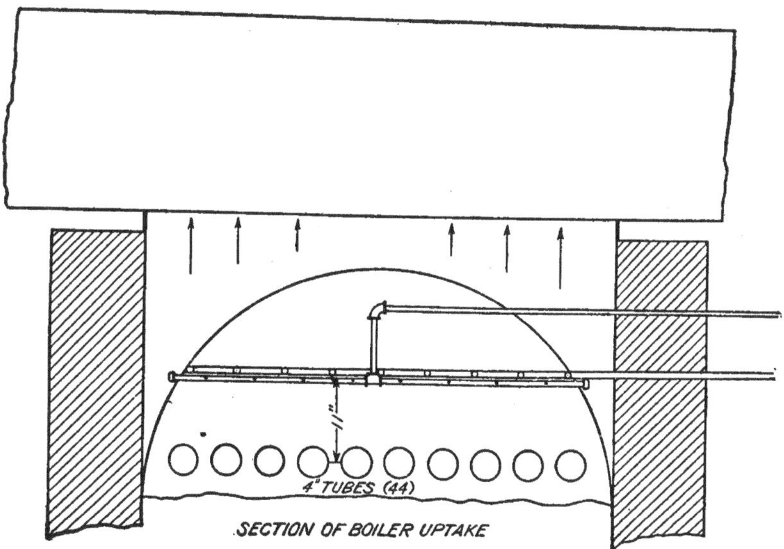

FIGURE 34.—Placing of nine open-end sampling tubes and a perforated pipe sampler in a horizontal return tubular boiler. The open-end tubes are of $\frac{3}{16}$-inch copper tubing.

EXPERIMENTS ON COLLECTING GASES.

In sampling flue gases it is desirable to collect the sample at a uniform rate for a period ranging from 15 minutes to several hours. The longer this period the more difficult it is to obtain a uniform rate of collecting gases with the ordinary 2-gallon collecting bottles. The main difficulty in collecting gas over long periods is that the water must be run out of the bottle through a very small opening which is apt to become stopped with small particles of dirt floating in the water. When the small opening is made by pinching the rubber tubing by the screw clamp shown in figure 9, the opening has a shape of a long and narrow slit which acts as a filter and catches all the dirt and becomes stopped. Better results can be obtained by using a small orifice like that shown in figure 35.

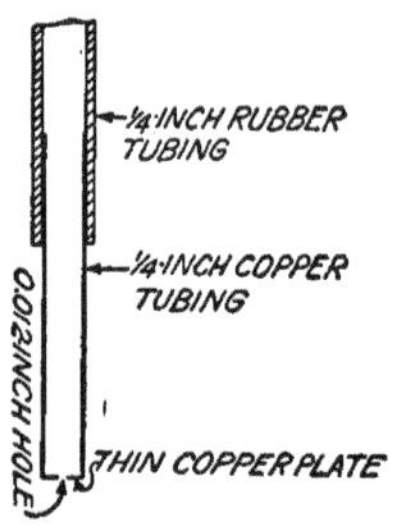

FIGURE 35.—Orifice used to obtain a uniform flow of water. The copper tube is inserted into the rubber tubing which discharges water into lower collecting bottle.

The orifice has a round opening which does not clog as easily as the long narrow slit formed by compressing the rubber tubing with the screw clamp. When collecting gas samples over long periods it is important that the water or brine used be free from all dirt. Filtering the brine or water through filter or blotter paper reduces the trouble from clogging.

Numerous experiments were made in the gas laboratory of the Bureau of Mines to determine how uniform a rate of collecting the gas may be expected when the sampling is extended over periods of 1 to 5 hours. Some of the results obtained are plotted in figure 36.

The horizontal distance of the chart gives the time over which the gas was collected. The vertical distance of the lower curves gives the volume of gas collected in a given period, each curve giving the volume with one adjustment of the screw clamp or an orifice. The vertical distance of the upper curves is the effective head in inches of water, this effective head decreasing as the water runs out of the bottle. The decrease in head gives a slight curvature to the lower curves, showing that the rate of collecting the gas is not quite uniform. It will be noticed that the curves obtained with the orifice are nearer a straight line than those obtained with the screw clamp. Curve 2 is a typical curve showing the results of dirt accumulating in the narrow slit in the rubber tubing and gradually stopping the flow of water. When the gas is collected over a period of 5 hours the water flows out of the collecting bottle at the rate of about two drops per second.

If the collection is started with an effective head smaller than those shown in figure 36, the rate of collection will not be as uniform as shown by the curve in the figure.

EXPERIMENTAL DATA ON THE ABSORPTION OF CO_2 BY WATER AND BRINE.

Gases are absorbed by liquids from a mixture of gases according to their partial perssures. The amount depends upon the nature

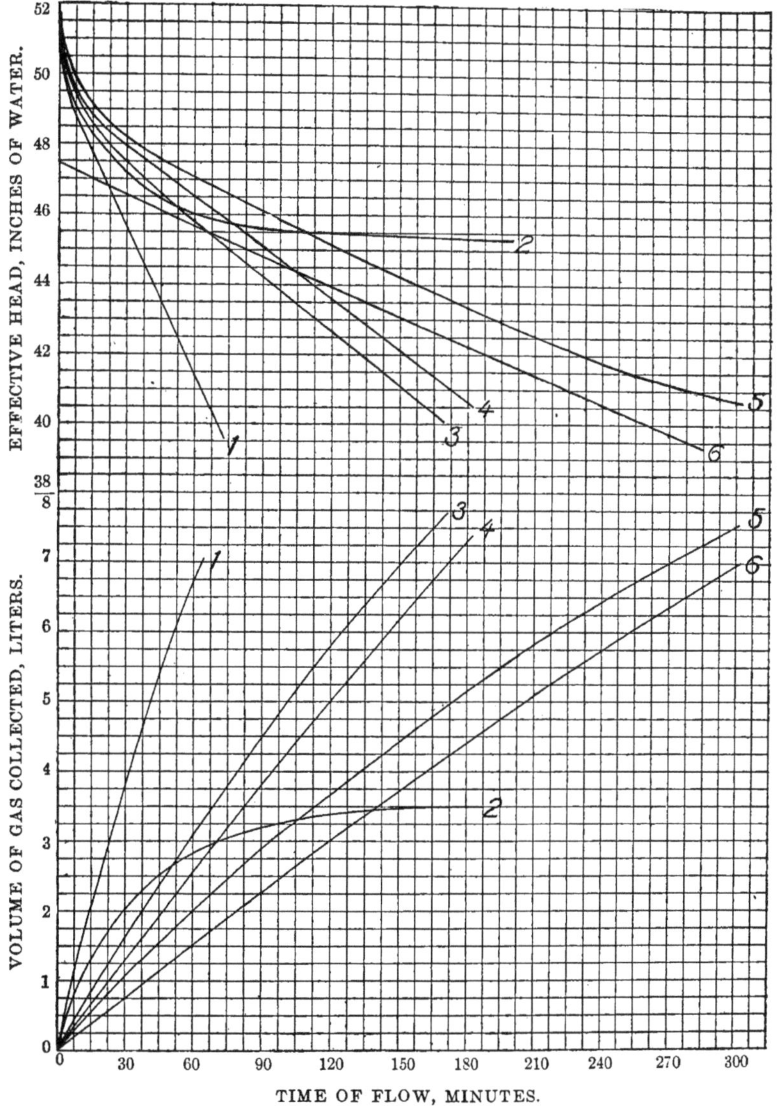

FIGURE 36.—Curves showing the uniformity of rate of collecting gas with two collecting bottles arranged as shown in figure 8. Upper curves show the change in effective head, lower curves the volume of gas collected. Curves 4 and 6 were obtained with orifice; all others with screw clamp.

of the substances and the temperature. Solids dissolved in liquids decrease the solubility of gases in the liquids, if there is no chemical action.

When flue-gas samples are being collected, the solubility of CO_2 in the liquid over which the gas is collected changes when the composition of the flue gas changes. If the liquid has been saturated with flue gas containing 10 per cent of CO_2, and a gas containing 4 per cent of CO_2 is collected over the liquid, CO_2 will pass from the liquid into the gas, and the percentage of CO_2 in the gas will increase. In a similar manner, if gas containing 10 per cent of CO_2 is collected over a liquid in equilibrium with 4 per cent of CO_2, some CO_2 will pass from the gas to the liquid and the percentage of CO_2 in the gas will decrease. These facts are shown clearly by the experimental results given in Tables 6 to 10.

The experiments were made to determine the extent to which these changes take place when water or a salt solution is used for the confining liquid.

The tests were all made under similar conditions, the same relative quantities of gas and liquid being used in each test. A change in the quantity of either will change the results, especially if the gas and the liquid are shaken together. The larger the contact surface between gas and liquid the greater will be the speed of interchange of CO_2.

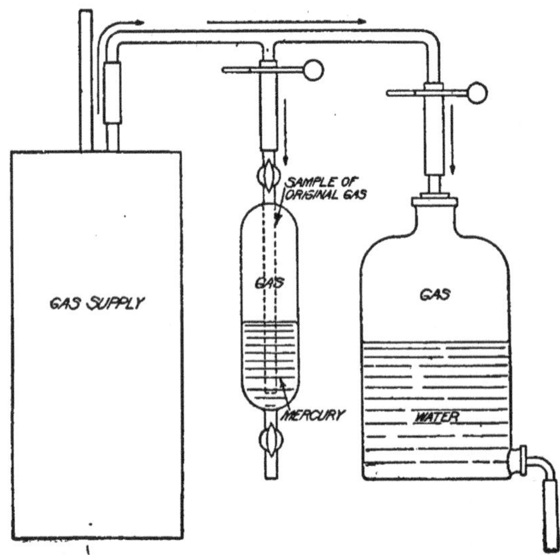

FIGURE 37.—Arrangement of apparatus for determining the absorption of CO_2 when collecting the gas over water or brine.

The experiments were made as follows: A 2-gallon bottle was completely filled with water or brine. About 1 liter of gas with a given percentage of CO_2 was then collected over the water. The water and the gas were shaken 5 minutes, then left to stand 30 minutes and again shaken for 5 minutes. With this procedure the water or the brine was brought nearly into equilibrium with the gas above it. A sample of the gas was analyzed for CO_2, and the figures obtained are the values given in the first column of the tables. The water or brine thus prepared was used in collecting a sample of gas containing either a lower or a higher percentage of CO_2 than the gas with which the liquid was previously in contact. As the gas was collected in the bottle a small part of the gas stream was led into a small gas container and collected over mercury. The arrangement of apparatus for the collection of these samples is shown in figure 37.

The collecting bottle was filled about two-thirds with gas and left standing undisturbed. The gas collected in the small gas container was of the same composition as the gas flowing into the large collecting bottle, and its composition was not affected by the mercury. Therefore its analysis gave the percentage of CO_2 in the gas as it flowed into the collecting bottle. In Tables 6, 8, and 10 this percentage is given in the third column, and in Tables 7 and 9, in the second column.

After each of 2, 4, and 21 hour periods a small gas sample was taken out of the large collecting bottle and analyzed. The percentages of CO_2 found by the analyses are given in Tables 6 to 10. In some experiments the temperature of the water in the collecting bottle was measured at the time a sample was taken for analysis. These temperature readings are given alongside the CO_2 percentages.

In some of the experiments after the gas and liquid had stood 21 hours and the percentage of CO_2 in the gas had been determined, the gas and liquid were shaken together for five minutes and the gas was again analyzed for CO_2. The results of such determinations of CO_2 are given in the second column of Tables 6 and 8.

The experiments were made with water with a 10 per cent brine solution and with saturated, or about 35 per cent, brine. The 10 per cent brine solution was prepared by dissolving 1 pound of common salt in 9 pounds of water. The saturated, or about 35 per cent, brine was obtained by dissolving all the salt that would go into solution—about 3½ pounds of salt per 10 pounds of water.

Table 9 gives the data on the absorption of CO_2 when the gas was bubbled through the liquid, with a constant effective head, as shown in figure 10. The duration of bubbling the gas through the liquid is given by the last column of the table.

Table 10 gives the data from experiments made to determine the effect of shaking the gas and water or brine together. Gas of given composition was collected over the liquid in a bottle, the collection lasting about five minutes. One sample was taken and analyzed immediately after collection. The bottle was then shaken for five minutes and the sample again analyzed for CO_2.

In all the experiments the same quantity of gas and liquid was used; that is, two-thirds of the volume of the bottle was gas and one-third was water.

TABLE 6.—*Data showing the tendency of water to absorb and give out CO_2.*

Percentage of CO_2 in gas after contact with water.	Temperature of water.	Percentage of CO_2 in original gas.	Percentage of CO_2 in gas after standing over water for— Two hours.	Temperature of water.[a]	Four hours.	Temperature of water.[a]	Twenty-one hours.	Temperature of water.[a]	Percentage of CO_2 in gas after standing 21 hours and being shaken 5 minutes.	Temperature of water when gas was analyzed.[b]
	°C.			°C.		°C.		°C.		°C.
5.0	....	8.9	8.8	....	8.6	....	7.9	....	....	....
5.0	....	11.6	11.3	....	11.0	....	9.2	....	....	....
8.0	....	11.9	11.7	....	11.8	....	10.9	....	....	....
4.0	....	15.0	[b] 14.9	....	13.6	....	10.5	....	....	....
7.7	....	4.4	4.6	....	4.8	....	5.4	....	....	....
17.0	....	2.7	3.1	....	3.5	....	6.3	....	....	....
4.5	21	10.3	10.1	22	9.9	23	8.8	22	8.0	22
4.6	29	10.8	10.8	30	10.1	32	9.3	27	8.6	27
5.0	23	13.0	12.9	24	12.6	25	10.6	20	9.2	20
6.6	27	8.8	8.7	29	8.5	29	7.9	27	7.9	27
10.1	28	3.9	4.2	30	4.4	31	5.2	22	5.9	22
12.0	24	5.6	5.7	24	5.9	25	6.7	26	8.3	26

[a] Temperature was recorded only where given in the table.
[b] Gas had stood over water only one hour.

TABLE 7.—*Data showing the tendency of 10 per cent brine to absorb and give out CO_2.*

Percentage of CO_2 in gas after contact with 10 per cent brine.	Percentage of CO_2 in original gas.	Percentage of CO_2 in gas after standing over 10 per cent brine for— Two hours.	Four hours.	Twenty-one hours.
3.9	8.9	8.9	9.0	8.2
4.0	15.0	[a] 15.0	14.5	12.7
5.0	11.6	11.5	11.5	10.2
8.0	11.9	11.8	11.6	11.0
10.6	4.4	4.5	4.6	6.7
17.0	2.7	2.7	3.1	4.9

[a] Gas had stood over brine only one hour.

TABLE 8.—*Data showing the tendency of saturated brine to absorb and give out CO_2.*

Percentage of CO_2 in gas after contact with saturated brine.	Temperature of water.	Percentage of CO_2 in original gas.	Percentage of CO_2 in gas after standing over saturated brine for— Two hours.	Temperature of water.	Four hours.	Temperature of water.	Twenty-one hours.	Temperature of water.	Percentage of CO_2 in gas after standing 21 hours over brine and then shaking for 5 minutes.	Temperature of water.
	°C.			°C.		°C.		°C.		°C.
2.5	21	10.3	10.3	22	10.2	23	9.7	22	8.1	22
5.5	23	13.0	13.1	24	13.0	25	12.0	20	11.3	20
10.4	28	3.9	4.0	30	4.1	31	4.4	22	4.5	22
12.0	24	5.6	5.8	24	5.6	25	5.9	26	6.5	26

TABLE 9.—*Data showing the effect of bubbling gas through water and brine with a constant head.*[a]

Percentage of CO_2 in gas after contact with brine.	Percentage of CO_2 in original gas.	Percentage of CO_2 after bubbling through water.	Percentage of CO_2 after bubbling through 10 per cent brine.	Time of collecting.
				Hours.
0.0	8.3	4.4	5.3	3½
4.5	13.2	7.6	8.3	3½
16.5	10.6	12.8	12.6	4
9.7	6.0	7.5	6.9	1

[a] See fig. 10 (p. 17).

TABLE 10.—*Data showing effect of shaking gas and water, or 35 per cent brine, in collecting bottle.*

Liquid shaken with gas.	Percentage of CO_2 in gas after contact with liquid.	Percentage of CO_2 in original gas.	Percentage of CO_2 after shaking gas and liquid five minutes.
Water	12.0	6.3	7.5
	7.4	11.7	10.8
	3.9	18.5	14.8
Saturated brine	11.9	6.2	7.3
	7.4	11.7	11.0
	3.9	18.5	16.7

CONCLUSIONS FROM EXPERIMENTS ON ABSORPTION OF CO_2 BY LIQUIDS.

The results of the experiments just described permit the following conclusion: With greatly varying compositions of flue gas the error resulting from absorption or giving out of CO_2 by the confining liquid may amount to 3 or 4 per cent of CO_2. The error is less with 10 per cent brine than with water and less with saturated brine than with 10 per cent brine.

Shaking the gas and the liquid together increases the absorption or giving out of CO_2 by the liquid.

The largest error is introduced when the gas is bubbled through water. Therefore this method of collecting gas should not be used.

If the percentage of CO_2 does not vary more than 2 or 3 per cent and if the gas does not stand over the liquid more than 2 hours, water may be used satisfactorily for the confining liquid. If, however, the variation of CO_2 exceeds 3 per cent and the gas stands over the liquid a long time, brine can be used to good advantage. The brine is not as convenient to handle as pure water; the water evaporates and the salt is deposited on all surfaces and mars the appearance of the apparatus.

Gas should be analyzed as soon as collected; shaking of the liquid and any change in its temperature should be avoided.

Ordinarily when collecting flue gases in the boiler room it is doubtful whether the percentage of CO_2 of the collected gas when analyzed is closer than 0.5 per cent of the gas as it enters the collecting bottle.

Considering all the possible errors that may and often do enter into the sampling and collecting of flue gas, one is justified in looking with suspicion on many of the published heat balances of boiler trials. The possible errors make the determination of radiation losses by difference in the heat balance questionable.

PUBLICATIONS ON FUEL TECHNOLOGY.

A limited supply of the following publications of the Bureau of Mines is temporarily available for free distribution. Requests for all publications can not be granted, and applicants should limit their selecting to publications that may be of especial interest to them. Requests for publications should be addressed to the Director, Bureau of Mines, Washington, D. C.

BULLETIN 1. The volatile matter of coal, by H. C. Porter and F. K. Ovitz. 1910. 56 pp., 1 pl., 9 figs.

BULLETIN 2. North Dakota lignite as a fuel for power-plant boilers, by D. T. Randall and Henry Kreisinger. 1910. 42 pp., 1 pl., 7 figs.

BULLETIN 3. The coke industry of the United States as related to the foundry, by Richard Moldenke. 1910. 32 pp.

BULLETIN 4. Features of producer-gas power-plant development in Europe, by R. H. Fernald. 1910. 27 pp., 4 pls., 7 figs.

BULLETIN 5. Washing and coking tests of coal at Denver, Colo., July 1, 1908, to June 30, 1909, by A. W. Belden, G. R. Delamater, J. J. Groves, and K. M. Way. 1910. 62 pp., 1 fig.

BULLETIN 6. Coals available for the manufacture of illuminating gas, by A. H. White and Perry Barker, compiled and revised by H. M. Wilson. 1911. 77 pp., 4 pls., 12 figs.

BULLETIN 7. Essential factors in the formation of producer gas, by J. K. Clement, L. H. Adams, and C. N. Haskins. 1911. 58 pp., 1 pl., 16 figs.

BULLETIN 12. Apparatus and methods for the sampling and analysis of furnace gases, by J. C. W. Frazer and E. J. Hoffman. 1911. 22 pp., 6 figs.

BULLETIN 13. Résumé of producer-gas investigations, October 1, 1904, to June 30, 1910, by R. H. Fernald and C. D. Smith. 1911. 393 pp., 12 pls., 250 figs.

BULLETIN 14. Briquetting tests of lignite at Pittsburgh, Pa., 1908–9; with a chapter on sulphite-pitch binder, by C. L. Wright. 1911. 64 pp., 11 pls., 4 figs.

BULLETIN 16. The uses of peat for fuel and other purposes, by C. A. Davis. 1911. 214 pp., 1 pl., 1 fig.

BULLETIN 18. The transmission of heat into steam boilers, by Henry Kreisinger and W. T. Ray. 1912. 180 pp., 78 figs.

BULLETIN 23. Steaming tests of coals and related investigations, September 1, 1904, to December 31, 1908, by L. P. Breckenridge, Henry Kreisinger, and W. T. Ray. 1912. 380 pp., 2 pls., 94 figs.

BULLETIN 24. Binders for coal briquets, by J. E. Mills. 56 pp., 1 fig. Reprint of United States Geological Survey Bulletin 343.

BULLETIN 27. Tests of coal and briquets as fuel for house-heating boilers, by D. T. Randall. 44 pp., 3 pls., 2 figs. Reprint of United States Geological Survey Bulletin 366.

BULLETIN 28. Experimental work conducted in the chemical laboratory of the United States fuel-tesing plant at St. Louis, Mo., January 1, 1905, to July 31, 1906, by N. W. Lord. 51 pp. Reprint of United States Geological Survey Bulletin 323.

BULLETIN 29. The effect of oxygen in coal, by David White. 80 pp., 3 pls. Reprint of United States Geological Survey Bulletin 382.

BULLETIN 31. Incidental problems in gas-producer tests, by R. H. Fernald, C. D. Smith, J. K. Clement, and H. A. Grine. 29 pp., 8 figs. Reprint of United States Geological Survey Bulletin 393.

BULLETIN 32. Commercial deductions from comparisons of gasoline and alcohol tests of internal-combustion engines, by R. M. Strong. 38 pp. Reprint of United States Geological Survey Bulletin 392.

BULLETIN 33. Comparative tests of run-of-mine and briquetted coal on the torpedo boat *Biddle*, by W. T. Ray and Henry Kreisinger. 50 pp., 10 figs. Reprint of United States Geological Survey Bulletin 403.

BULLETIN 34. Tests of run-of-mine and briquetted coal in a locomotive boiler, by W. T. Ray and Henry Kreisinger. 33 pp., 9 figs. Reprint of United States Geological Survey Bulletin 412.

BULLETIN 35. The utilization of fuel in locomotive practice, by W. F. M. Goss. 29 pp., 8 figs. Reprint of United States Geological Survey Bulletin 402.

BULLETIN 36. Alaskan coal problems, by W. L. Fisher. 1911. 32 pp., 1 pl.

BULLETIN 39. The smoke problem at boiler plants, a preliminary report, by D. T. Randall. 31 pp. Reprint of United States Geological Survey Bulletin 334, revised by S. B. Flagg.

BULLETIN 40. The smokeless combustion of coal in boiler furnaces, with a chapter on central heating plants, by D. T. Randall and H. W. Weeks. 188 pp., 40 figs. Reprint of United States Geological Survey Bulletin 373, revised by Henry Kreisinger.

BULLETIN 43. Comparative fuel values of gasoline and denatured alcohol in internal-combustion engines, by R. M. Strong and Lauson Stone. 1912. 243 pp., 3 pls., 32 figs.

BULLETIN 49. City smoke ordinances and smoke abatement, by S. B. Flagg. 1912. 55 pp.

BULLETIN 54. Foundry cupola gases and temperatures, by A. W. Belden. 1913. 29 pp., 3 pls., 16 figs.

BULLETIN 55. The commercial trend of the producer-gas power plant in the United States, by R. H. Fernald. 1913. 93 pp., 1 pl., 4 figs.

BULLETIN 56. First series of coal-dust tests in the experimental mine, by G. S. Rice, L. M. Jones, J. K. Clement, and W. L. Egy. 1913. 115 pp., 12 pls., 28 figs.

BULLETIN 58. Fuel-briquetting investigations, July, 1904, to July, 1912, by C. L. Wright. 1913. 277 pp., 21 pls., 3 figs.

BULLETIN 76. United States coals available for export trade, by Van. H. Manning. 1914. 15 pp., 1 pl.

BULLETIN 88. The condensation of gasoline from natural gas, by G. A. Burrell, F. M. Seibert, and G. G. Oberfell. 1915. 106 pp., 6 pls., 18 figs.

TECHNICAL PAPER 1. The sampling of coal in the mine, by J. A. Holmes. 1911. 18 pp., 1 fig.

TECHNICAL PAPER 2. The escape of gas from coal, by H. C. Porter and F. K. Ovitz. 1911. 14 pp., 1 fig.

TECHNICAL PAPER 3. Specifications for the purchase of fuel oil for the Government, with directions for sampling oil and natural gas, by I. C. Allen. 1911. 13 pp.

TECHNICAL PAPER 5. The constituents of coal soluble in phenol, by J. C. W. Frazer and E. J. Hoffman. 1912. 20 pp., 1 pl.

TECHNICAL PAPER 8. Methods of analyzing coal and coke, by F. M. Stanton and A. C. Fieldner. 1913. 42 pp., 12 figs.

TECHNICAL PAPER 10. Liquefied products of natural gas; their properties and uses, by I. C. Allen and G. A. Burrell. 1912. 23 pp.

TECHNICAL PAPER 16. Deterioration and spontaneous combustion of coal in storage, a preliminary report, by H. C. Porter and F. K. Ovitz. 1912. 14 pp.

TECHNICAL PAPER 20. The slagging type of gas producer, with a brief report of preliminary tests, by C. D. Smith. 1912. 14 pp., 1 pl.

TECHNICAL PAPER 25. Methods for the determination of water in petroleum and its products, by I. C. Allen and W. A. Jacobs. 1912. 13 pp., 2 figs.

TECHNICAL PAPER 31. Apparatus for the exact analysis of flue gas, by G. A. Burrell and F. M. Seibert. 1913. 12 pp., 1 fig.

TECHNICAL PAPER 34. Experiments with furnaces for a hand-fired return tubular boiler, by S. B. Flagg, G. C. Cook, and F. E. Woodman. 1914. 32 pp., 1 pl., 4 figs.

TECHNICAL PAPER 35. Weathering of the Pittsburgh coal bed at the experimental mine near Bruceton, Pa., by H. C. Porter and A. C. Fieldner. 1914. 35 pp., 14 figs.

TECHNICAL PAPER 37. Heavy oil as fuel for internal-combustion engines, by I. C. Allen. 1913. 36 pp.

TECHNICAL PAPER 38. Wastes in the production and utilization of natural gas, and methods for their prevention, by Ralph Arnold and F. G. Clapp. 1913. 29 pp.

TECHNICAL PAPER 45. Waste of oil and gas in the Mid-Continent fields, by R. S. Blatchley. 1914. 54 pp., 2 pls., 15 figs.

TECHNICAL PAPER 49. The flash point of oils, methods and apparatus for its determination, by I. C. Allen and A. S. Crossfield. 1913. 31 pp., 2 figs.

TECHNICAL PAPER 50. Metallurgical coke, by A. W. Belden. 1913. 48 pp., 1 pl., 23 figs.

TECHNICAL PAPER 55. The production and use of brown coal in the vicinity of Cologne, Germany, by C. A. Davis. 1913. 15 pp.

TECHNICAL PAPER 57. A preliminary report on the utilization of petroleum and natural gas in Wyoming, by W. R. Calvert, with a discussion of the suitability of natural gas for making gasoline, by G. A. Burrell. 1913. 23 pp.

TECHNICAL PAPER 63. Factors governing the combustion of coal in boiler furnaces; a preliminary report, by J. K. Clement, J. C. W. Frazer, and C. E. Augustine. 1914. 46 pp., 26 figs.

TECHNICAL PAPER 65. A study of the oxidation of coal, by H. C. Porter. 1914. 30 pp., 12 figs.

TECHNICAL PAPER 76. Notes on the sampling and analysis of coal, by A. C. Fieldner. 1914. 59 pp., 6 figs.

TECHNICAL PAPER 80. Hand firing soft coal under power plant boilers, by H. Kreisinger. 1915. 83 pp., 32 figs.

TECHNICAL PAPER 89. Coal products, and the possibility of their successful manufacture, by H. C. Porter, with a chapter on coal-tar products used in explosives, by C. G. Storm. 1915. 21 pp.

TECHNICAL PAPER 109. Composition of the natural gas used in 25 cities, with a discussion of the properties of natural gas, by G. A. Burrell and G. G. Oberfell. 1915. 22 pp.

TECHNICAL PAPER 112. The explosibility of acetylene, by G. A. Burrell and G. G. Oberfell. 1915. 15 pp.

TECHNICAL PAPER 115. Inflammability of mixtures of gasoline vapor and air, by G. A. Burrell and H. T. Boyd. 1915. 18 pp., 2 figs.

TECHNICAL PAPER 120. A bibliography of the chemistry of gas manufacture, by W. H. Rittman and C. F. Whittaker. 1915. 37 pp.

TECHNICAL PAPER 123. Notes on the use of low-grade fuels in Europe, by R. H. Fernald. 1915. 37 pp., 4 pls., 4 figs.

INDEX.

CPSIA information can be obtained
at www.ICGtesting.com
Printed in the USA
BVHW091423211118
533722BV00029B/2007/P